T0178592

Green Energy and Technology

Climate change, environmental impact and the limited natural resources urge scientific research and novel technical solutions. The monograph series Green Energy and Technology serves as a publishing platform for scientific and technological approaches to "green"—i.e. environmentally friendly and sustainable—technologies. While a focus lies on energy and power supply, it also covers "green" solutions in industrial engineering and engineering design. Green Energy and Technology addresses researchers, advanced students, technical consultants as well as decision makers in industries and politics. Hence, the level of presentation spans from instructional to highly technical.

Indexed in Scopus.

More information about this series at http://www.springer.com/series/8059

Phaik Eong Poh · Ta Yeong Wu ·
Weng Hoong Lam · Wai Ching Poon ·
Chean Shen Lim

Waste Management in the Palm Oil Industry

Plantation and Milling Processes

 Springer

Phaik Eong Poh
School of Engineering
Monash University Malaysia
Bandar Sunway, Selangor, Malaysia

Weng Hoong Lam
School of Energy and Chemical Engineering
Xiamen University Malaysia
Bandar Sunsuria, Selangor, Malaysia

College of Chemistry and Chemical
Engineering
Xiamen University
Xiamen, China

Chean Shen Lim
Wings Strategic Management Sdn Bhd
Shah Alam, Selangor, Malaysia

Ta Yeong Wu
School of Engineering
Monash University Malaysia
Bandar Sunway, Selangor, Malaysia

Wai Ching Poon
School of Business
Monash University Malaysia
Bandar Sunway, Selangor, Malaysia

ISSN 1865-3529   ISSN 1865-3537   (electronic)
Green Energy and Technology
ISBN 978-3-030-39552-0   ISBN 978-3-030-39550-6   (eBook)
https://doi.org/10.1007/978-3-030-39550-6

This Springer imprint is published by the registered company Springer Nature Switzerland AG
The registered company address is: Gewerbestrasse 11, 6330 Cham, Switzerland

Preface

Countless literature published in the form of journal articles, reviews, book chapters, and conference proceedings on palm oil wastes can be found these days. These are mostly accessible to scientists and researchers in the field, and not so for practitioners in the palm oil industry, especially those who are new to the industry and wish to learn more.

This book aims to provide readers an overview of palm oil activities, covering plantation and milling process, and issues that were raised in the past and present with regards to the sustainability of these activities. Also, the focus is given to waste generation from plantation and milling in the palm oil sector and how these wastes could be managed effectively from a technological viewpoint.

Readers can also refer to this book for a case study on the introduction of a business scheme to convert mill wastes into bioorganic fertilizer. We evaluate the economic, social, and environmental benefits of such a program and discuss future challenges from sustainability, industry's viewpoint as well as government policies. This book will enable practitioners to effectively review and decide on projects that can elevate the overall sustainability of the palm oil industry.

Bandar Sunway, Malaysia   Phaik Eong Poh
Bandar Sunway, Malaysia   Ta Yeong Wu
Bandar Sunsuria, Malaysia/Xiamen, China   Weng Hoong Lam
Bandar Sunway, Malaysia   Wai Ching Poon
Shah Alam, Malaysia   Chean Shen Lim

Contents

Chapter 1
Introduction

Oil palm tree (*Elaeis guineensis*), which originates from Africa, was brought to Malaysia initially as an ornamental plant by the British. In the 60s, the cultivation of oil palm in the country was intensified to reduce economic reliance from the tin and rubber industry. Currently, the oil palm industry is dominated by Southeast Asian countries such as Indonesia, Malaysia, and Thailand, wherein 2017, Indonesia and Malaysia alone contributed to 85% of the world's palm oil production (Iskandar et al. 2018).

Despite the lower fresh fruit bunch (FFB) yield in 2018, Malaysia's export of palm oil products remained strong, with an increase of 3.8% to 24.88 million tonnes (MPOB 2019). The versatility of palm oil to be converted into various products and its potential as an energy source is expected to lead to greater demand in the near future. However, the development of the oil palm industry has been clouded by controversies on the topic of sustainability and environmental impacts attributed to activities by the industry.

1.1 Deforestation

One hugely debated controversies related to the palm oil industry is deforestation. While oil palm fruits are the most efficient oil-yielding crop, requiring only 0.26 ha of land per tonne of oil as compared to other food crops such as sunflower, soybean, and rapeseed, a large area of rainforest or peatland were displaced to allow cultivation of oil palm to cope with the increasing demand. It was reported that new plantations developed between 1990 and 2010 in Indonesia were predominantly (63%) on deforested lands, while Malaysia had 17% of new estates located on deforested areas (Gunarso et al. 2013; Koh et al. 2011).

The impact of deforestation is, in fact, huge with disruption to the forest ecosystem, increased anthropogenic greenhouse gas emissions, and pollution to watercourse just to name a few. With the increasing demand for palm oil for various end products, the impact of the rapid growth of oil palm plantations is catching the attention of the

© Springer Nature Switzerland AG 2020
P. E. Poh et al., *Waste Management in the Palm Oil Industry*,
Green Energy and Technology, https://doi.org/10.1007/978-3-030-39550-6_1

world. More recently, transboundary haze is a widespread issue in ASEAN countries due to the clearing of forests to make way for oil palm plantations, where the majority of the forest fires occur in Indonesia (Jones 2006).

1.2   What Are the Major Countries Doing to Reduce the Impact Caused by the Oil Palm Industry?

While the industry has been contributing to the economic growth of many nations, the environmental and social impact that is related to the industry is not possible to be ignored. In view of the effects caused by the palm oil industry, Roundtable for Sustainable Palm Oil (RSPO) is a global establishment set up in 2004 to encourage sustainable development and utilization of palm oil products in accordance to a global standard, with the involvement of stakeholders (Iskandar et al. 2018). This organization involved stakeholders from varied backgrounds, from upstream to downstream of the palm oil process chain, retailers, NGOs, financial institutions of various countries with active production and usage of palm oil.

RSPO has developed a set of environmental and social criteria which enables companies to minimize negative impacts of palm oil cultivation on communities and environment in regions involved, producing Certified Sustainable Palm Oil (CSPO). However, only 19% of the total palm oil produced globally is certified by RSPO (RSPO 2019) and most of the CSPO producers are from the significant market players in Indonesia and Malaysia. Despite the push toward sustainability, the uptake by various companies would be considered low and more effort has to be made locally to encourage the shift toward the production of sustainable palm oil.

1.2.1   Indonesia, Malaysia, and Thailand

As Indonesia produces the primary volume of palm oil for local consumption, the country has a program—Indonesia Sustainable Palm Oil (ISPO) program, that is both mandatory and legally binding for all palm oil growers, with the exception of domestic smallholders. The aims of ISPO are very much similar to RSPO, however, the palm oil growers are subjected to the Presidential and Ministry of Agriculture Regulation. Meanwhile, export activities of CPO are regulated under the Ministry of Trade of the Republic of Indonesia (Andoko and Zmudczynska 2019).

While Malaysia has Malaysian Palm Oil Board (MPOB) as a government agency and Malaysian Palm Oil Council (MPOC) that both focuses on the development and market expansion of palm oil industry, Malaysia also has Malaysia Sustainable Palm Oil (MSPO) certification scheme that is managed by Malaysia Palm Oil Certification Council (MPOCC). MSPO used to be a body which provides certification and prescribes standards in oil palm management and supply chain and is not mandatory until recently. Plantation industries that already have RSPO certification has to fulfil MSPO standards by 31st December 2018 while those that are not RSPO certified were required to obtain certification by 30th June 2019. Meanwhile, smalholders have an extended timeline to meet MSPO standards by 31st December 2019. Having MSPO certification indicates that the company follows the standard and is committed to producing CSPO, ensuring the sustainability of the industry. Meanwhile, Thailand has developed Good Agricultural Practices standard (Thai Gap) in 2010 with the aim of monitoring issues in the palm oil industry to reduce the environmental impact of the sector (Iskandar et al. 2018).

1.3   What About Waste Materials from Processes?

The earlier discussions were surrounding the sustainability of the palm oil industry in terms of the environmental and social impact of expanding oil palm plantations. Mitigation plans can be put in place to improve the sustainability of plantations. Nevertheless, it is also crucial to focus on the wastes that are generated in the process of producing crude palm oil (CPO). This would cover the wastes generated from oil palm plantations such as oil palm fronds (OPF) and oil palm trunks (OPT), to solid and liquid wastes generated from the palm oil milling process which will be further elaborated in the next two chapters.

While it is important to reduce deforestation and mitigate the impact of change to the ecosystem, wastes generated from different processes to produce CPO are equally detrimental if not adequately monitored and regulated. It can lead to contamination of watercourse, air pollution and generation of uncaptured greenhouse gases. In addition, some of these waste materials have the potential to be harnessed into value-added product, that could also contribute to the expansion of palm oil industry or even the possibility of linking the different process chains to a closed loop, minimizing the generation of wastes and increasing the value chain of the entire industry.

In fact, the mills and plantations have been looking at various strategies to minimize waste production for a more sustainable process. Likewise, a lot of research and development work has been conducted in this field and some of the research output has been successfully implemented in the field. Nevertheless, only a handful of research outcomes has made its way to be known by professionals in the field, and the same goes for some successful stories in the field that are not widely publicized. Therefore, it is our aim to compile these informations together and critically analyze the different solutions available, providing suggestions to the way forward to build a more sustainable palm oil industry.

As such, this book is categorized into four major chapters: (1) giving focus to solid waste materials that are generated in the plantation, and their potential to be converted to valuable raw materials for subsequent downstream processes; (2) the importance of managing solid and liquid wastes generated from palm oil mills, alternatives to deal with them, and potential of these wastes to generate renewable energy; (3) a zoom in to technological improvements of anaerobic digestion of palm oil mill effluent (POME); and finally (4) a case study on palm oil waste management initiatives—whether it is going to be sustainable.

References

Andoko, E., & Zmudczynska, E. (2019). An analysis of the palm industry's development, regulations, and practices in Indonesia. *FFTC Agricultural Policy Platform*. Retrieved November 26, 2019, from http://ap.fftc.agnet.org/ap_db.php?id=996.

Gunarso, P., Hartoyo, M. E., Agus, F., & Killeen, T. J. (2013). Oil palm and land use change in Indonesia, Malaysia and Papua New Guinea. Reports from the Technical Panels of the 2nd Greenhouse Gas Working Group of the Roundtable on Sustainable Palm Oil (RSPO) (pp. 29–64).

Iskandar, M. J., Baharum, A., Anuar, F. H., & Othaman, R. (2018). Palm oil industry in South East Asia and the effluent treatment technology—A review. *Environmental Technology & Innovation, 9*, 169–185.

Jones, D. S. (2006). ASEAN and transboundary haze pollution in Southeast Asia. *Asia Europe Journal, 4*, 431–446.

Koh, L. P., Miettinen, J., Liew, S. C., & Ghazoul, J. (2011). Remotely sensed evidence of tropical peatland conversion to oil palm. *Proceedings of the National Academy of Sciences of the United States of America, 108*, 5127–5132.

Malaysian Palm Oil Board (MPOB). (2019). Overview of the Malaysian oil palm industry 2018. Malaysia Palm Oil Board (pp. 1–6).

Roundtable for Sustainable Palm Oil. (2019). RSPO—Roundtable for Sustainable Palm Oil. Retrieved November 26, 2019, from https://rspo.org/about.

Chapter 2
Oil Palm Plantation Wastes

2.1 Introduction

Biorefinery is one of the many sustainable approaches that emphasizes on converting the lignocellulosic biomass into a wide variety of valuable bio-based products (Chandel et al. 2018). The main driving force that leads to the rapid growth of biorefinery is due to high dependency on nonrenewable resources known as fossil fuels. Massive consumption of fossil fuels is found to be one of the major contributors to the emergence of environmental impacts such as global warming and climate change. Extensive researches were dedicated to searching the alternative to the crude oil refineries (Cherubini 2010). Among the numerous technologies, biorefinery has been identified as one of the highly promising alternatives to the conventional refineries.

In Malaysia, the palm oil industry is an irreplaceable economic activity due to its high global demand in both food and nonfood industries. Thus, an enormous oil palm plantation is necessary to fulfill the need for palm oil production. However, a large quantity of agriculture wastes is also generated in the oil palm plantation. Some of the common agriculture wastes found in oil palm plantation include oil palm fronds (OPF) and oil palm trunk (OPT). One hectare of oil palm planted area produces approximately 10.40 tonnes of OPF and 74.48 tonnes of OPT (Loh 2017). Both OPF and OPT are often referred to as oil palm plantation wastes. In fact, these oil palm wastes pose great potential to be reutilized as the lignocellulosic feedstocks for biorefineries.

Lignocellulosic biomass is a plant-based material consists of three key components, known as cellulose, hemicellulose, and lignin. In general, biorefineries of lignocellulosic biomass can be classified into two main routes, polysaccharides and lignin (Chandel et al. 2018). Obviously, the polysaccharides path utilizes cellulose or hemicellulose, while the lignin path utilizes only the lignin. When chemicals and/or heat are applied, hemicellulose is relatively easier to dissociate as compared to cellulose (Ho et al. 2019). A typical OPF contains approximately 44.46% of cellulose, 23.53% of hemicellulose, and 17.20% of lignin (Ho and Wu 2020). On the contrary,

© Springer Nature Switzerland AG 2020
P. E. Poh et al., *Waste Management in the Palm Oil Industry*,
Green Energy and Technology, https://doi.org/10.1007/978-3-030-39550-6_2

an OPT is composed of about 50.78%, 30.36% and 17.87% of cellulose, hemicellulose and lignin, respectively (Lai and Idris 2013). The compositions of OPF and OPT may vary, depending on the maturity and location source of the oil palm tree in which the wastes are obtained (Loow et al. 2017b, 2018; Ong et al. 2019). The relatively high composition of carbohydrates presents in both OPF and OPT indicates polysaccharides route is preferred for these oil palm wastes. Nevertheless, the presence of an appreciable amount of lignin denoted the potential use of OPF and OPT for the lignin route. This book chapter highlights the applications of OPF and OPT in multiple industries and fields, including bioenergy, bio-based chemicals, biochar, fertilizer, and animal feed.

2.2 Potential Application of Oil Palm Wastes

2.2.1 Bioenergy

The global market for biodiesel for 2019 is estimated to be US$ 35.4 billion (Ahmad et al. 2019). Lignocellulosic biomass from oil palm production has the potential to be converted into high-value bioproducts, especially biofuel (Ahmad et al. 2019). In this regard, OPF and OPT are valuable and cheap bioresources for the biorefinery industry to produce biofuels. Recently, Farah Amani et al. (2018) used OPF hydrolysate obtained from enzymatic hydrolysis as a substrate to produce bioethanol via *Saccharomyces cerevisiae* HC 10. Using the OPF hydrolysate, the biomass yield coefficient and bioethanol yield coefficient were 0.1623 g cell/g sugar and 0.1191 g bioethanol/g sugar, respectively. Additionally, the highest bioethanol concentration, up to 7.23 g/L, was detected after one day of the fermentation process (Farah Amani et al. 2018). Farah Amani et al. (2018) also found that the combinations of cellulase and hemicellulase enzymes produced higher concentration of reducing sugars as compared to cellulase or hemicellulase used individually. Mastuli et al. (2017) investigated the use of supercritical water gasification of OPF to produce biohydrogen. Oxide of non-noble metals such as NiO, CuO, and ZnO were impregnated onto the MgO surface at 20 wt% and used as catalysts for the supercritical water gasification processes. Their results showed that the 20ZnO/MgO catalyst exhibited the highest biohydrogen yield. From this study, Mastuli et al. (2017) concluded that other factors such as dispersion, basicity, and bond strength played very important roles for higher catalytic performances. Also, all the non-noble metal-supported catalysts showed higher biohydrogen selectivity. Levulinic acid is a versatile platform chemical that can be derived from biomass as an alternative to fossil fuel resources. Ramli and Amin (2017) reused OPF, which was catalyzed by an acidic ionic liquid (1-sulfonic acid-3-methyl imidazolium tetrachloroferrate), to produce levulinic acid. At optimum conditions, 69.2% of levulinic acid yield was obtained from glucose while 24.8% of levulinic acid yield was attained from OPF and 77.3% of process efficiency could be achieved. After the levulinic acid was generated at the optimum conditions of glucose and OPF conversions, levulinic acid in the reaction product could be completely converted to ethyl levulinate through esterification with ethanol over the catalyst (Ramli and Amin 2017). Additionally, the recycled catalyst gave sufficient performance up to five successive cycles.

Ang et al. (2015) reported the potential reuse of OPT as an alternative fermentation feedstock for lignocellulolytic enzyme production, and carbon source for bioethanol production by *Aspergillus fumigatus* SK1 and *Candida tropicalis* RETL-CrI, respectively. The enzymatic production study was done using solid-state fermentation, whereby a maximum xylanase activity of 1792.43 U/g could be obtained, together with CMCase (56.19 U/g), FPase (3.47 U/g), and β-glucosidase (1.55 U/g). The physiochemical analysis of the hydrolyzed OPT showed severe destruction of the fiber microstructure by those crude enzymes. On the other hand, alcoholic fermentation of OPT hydrolysate using *C. tropicalis* produced 3.067 g/L of bioethanol, with a theoretical bioethanol yield of 68.05%. According to Eom et al. (2015a, b), higher production of bioethanol could be achieved from dried OPT which was treated by hydrothermolysis and enzymatic hydrolysis. Firstly, OPT was subjected to hydrothermal treatment at 180 °C and 30 min. After enzymatic hydrolysis of pretreated whole slurry of OPT, 43.5 g of glucose per 100 g dry biomass could be obtained, corresponding to 81.3% of the theoretical glucose yield. Using subsequent alcohol fermentation, 81.4% bioethanol yield of the theoretical bioethanol yield was achieved. Eom et al. (2015a, b) then conducted the proposed new process, in which starch in OPT was converted to bioethanol through enzymatic hydrolysis and subsequent fermentation prior to hydrothermal treatment, and the resulting slurry was subjected to identical processes that were applied to a control. Consequently, a high glucose yield of 96.3% was achieved, and the resulting ethanol yield was 93.5%. This proposed new method by Eom et al. (2015a, b) offered the advantage of reducing operational and capital costs due to minimizing the process for bioethanol production by excluding expensive processes related to detoxification prior to enzymatic hydrolysis and fermentation. Mohd Zakria et al. (2017) tested the potential of using *Kluyveromyces marxianus* ATCC 46,537 to produce bioethanol using OPT sap. The results showed that *K. marxianus* could produce higher bioethanol concentration at 16 h as compared to *S. cerevisiae*. Also, they found that magnesium sulfate and β-alanine enhanced bioethanol production using OPT sap. For example, when 7.93 g/L of magnesium sulfate and 0.90 g/L of β-alanine were used together with OPT sap, bioethanol production was increased 20% with a bioethanol yield and a productivity of 0.47 g/g and 2.22 g/L h, respectively. This was because magnesium involved in many physiological functions, including growth, cell division, and enzyme activity while β-alanine was a building block of pantothenic acid, which protected the cells from product inhibition and increased the biomass yield (Mohd Zakria et al. 2017).

Bardant et al. (2017) found that using response surface methodology, bioethanol production could be enhanced and can reach 7.92%v by using 36.5%w of OPT but the results were averagely 2.46%v lower than palm oil empty fruit bunch. However, it can be assumed that the optimum condition for high-loading-substrate enzymatic hydrolysis of OPT which followed by unsterilized fermentation for bioethanol production was similar to the optimum condition for palm oil empty fruit bunch pulp. A utilization of mixed-culture yeast was also investigated to produce bioethanol from unsterilized hydrolysis product but the improvement was not significant as compared to single-culture yeast. This was because the measurement of dried residue of single culture fermentation and mixed-culture fermentation, which was mostly pulp residue, did not show significant difference (Bardant et al. 2017). Sitthikitpanya et al. (2017, 2018) explored the production of biohydrogen and biomethane from OPT using *Thermoanaerobacterium thermosaccharolyticum* KKU19. The two-stage process for thermophilic biohydrogen production, followed by biomethane production was examined by Sitthikitpanya et al. (2017) using a hydrolysate obtained from OPT. The optimum conditions for lime pretreatment and enzymatic hydrolysis of OPT were found to be a lime loading of 0.2 g $Ca(OH)_2$/g-OPT, pretreatment time of 60 min, temperature of 121 °C, and enzyme loading of 35 filter paper units/g-OPT. Then, the OPT hydrolysate was used as a substrate for biohydrogen production by *T. thermosaccharolyticum* in the first stage. The maximum biohydrogen production potential of 2179 mL $H_2/L_{substrate}$ was obtained. Then, the acidic effluent was used to produce the biomethane in the second stage. The two-stage biohydrogen and biomethane production produced an energy yield of 10.6 kJ/g-COD_{added} with up to 83% COD removal (Sitthikitpanya et al. 2017). Instead of using the hydrolysate obtained from OPT, Sitthikitpanya et al. (2018) used lime-pretreated OPT to produce biohydrogen and biomethane by simultaneous saccharification and fermentation. The study found that the pretreatment of OPT by lime and solid-state fermentation decreased the crystallinity of OPT. Substrate loading and initial pH were significant factors affecting biohydrogen production from lime-pretreated OPT by *T. thermosaccharolyticum* KKU19. The optimal solid-state fermentation conditions for maximizing biohydrogen production were substrate loading, enzyme loading, inoculum concentration, initial pH, and temperature of 4.6%, 10 FPU/g-OPT, 10% (v/v), 6.3, and 50 °C, respectively, in which a biohydrogen yield of 60.22 mL H_2/g-OPT could be obtained. A two-stage thermophilic biohydrogen and biomethane production process effectively enhanced the energy yield from lime-pretreated OPT by 18-fold over the energy yield obtained from a single-stage hydrogen production process (Sitthikitpanya et al. 2018).

2.2.2   Bio-Based Chemicals

Chemical product derived from a renewable resource is commonly termed as bio-based chemical. Biorefinery processing of lignocellulosic biomass into chemical is one of the most attractive applications for a given biomass. In Malaysia, valorization of oil palm wastes such as OPF and OPT into value-added chemicals is highly desired from the aspects of clean production and sustainable development. As mentioned, lignin, hemicellulose, and cellulose are the three major constituents available in any species of lignocellulosic biomass. Therefore, a distinctive range of bio-based chemicals can be selectively produced by targeting different constituents present in the oil palm wastes (Chandel et al. 2018).

(i) Conversion steps from oil palm wastes into bio-based chemical

In general, synthesis of bio-based chemicals using biomass as the feedstock begins with pretreatment, followed by hydrolysis and ultimately the catalytic reaction. Pretreatment is required for raw oil palm wastes due to the presence of lignin; this protective layer hinders the accessibility of hydrolysis attack toward the carbohydrates and consequently leads to poor sugar extraction yield (Brandt et al. 2013). In contrast, pretreated oil palm wastes are readily hydrolyzed for sugar recovery. The hydrolysis process results in the of carbohydrate polymers into their respective monomeric sugars (Kumar et al. 2017). Specifically, hemicellulose is degraded into pentose (xylose and arabinose) while cellulose is degraded into hexose (glucose). These monomeric sugars obtained from the carbohydrates are the key substrates for the synthesis of the desired chemical product in the subsequent conversion step. Finally, the conversion of monomeric sugars into bio-based chemicals is mainly undergone via either microbial fermentation or chemical modification (Cherubini 2010).

(ii) Applications of oil palm wastes in the production of potential bio-based chemicals

The potential platform bio-based chemicals proposed by the US Department of Energy (USDOE) based on green chemistry in the year 2010 are listed in Table 2.1. Platform chemical is an intermediate which can be further manufactured into a broad range of useful products. In the past decade, an enormous number of reactive systems has been developed for the production of platform chemicals using oil palm wastes as sustainable feedstocks. Table 2.2 summarized the production of different bio-based chemicals from OPF and OPT.

As seen in Table 2.2, the recent publications of bio-based chemicals produced from OPF and OPT are similar to the list of chemicals as shown in Table 2.1. Here, furfural and 2,3-butanediol are the members of furan and glycerol, respectively. This further indicates that the major trend of bio-based chemicals is strongly associated with the top platform chemicals suggested by USDOE. The overall conversion of

Table 2.1 Top platform chemicals proposed by USDOE (U.S. Department of Energy, 2011)	Top platform chemicals in the year 2010
	• Bioethanol
	• Furan and its derivatives
	• Glycerol and its derivatives
	• Hydrocarbons
	• Lactic acid
	• Succinic acid/Aldehyde/3-hydroxy propionic acid
	• Levulinic acid
	• Sorbitol
	• Xylitol

Table 2.2 Production of bio-based chemicals from OPF and OPT

Feedstock	Experimental conditions	Product	Product yield (%)	References
OPF	Temperature = 100 °C Time = 135 min Solvent = Choline chloride-oxalic acid	Furfural	26.34	Lee et al. (2019)
OPF	Temperature = 37 °C Time = 24 h Microbe = *Enterobacter cloacae* SG1	2,3-butanediol	25.56	Hazeena et al. (2016)
OPT sap	Temperature = 50 °C Time = 12 h Microbe = *Bacillus coagulans* strain 191	Lactic acid	92.00	Kunasundari et al. (2017)
OPT	Temperature = 37 °C Time = 28 h Microbe = *Lactobacillus paracasei*	Lactic acid	89.50	Eom et al. (2015a)
OPT	Temperature = 37 °C Dilution rate = 0.4/h Stabilization time = 634 h Microbe = *Actinobacillus succinogenes* 130Z	Succinic acid	69.00	Luthfi et al. (2019)
OPT	Temperature = 37 °C Time = 24 h Microbe = *Actinobacillus succinogenes* 130Z	Succinic acid	26.00	Bukhari et al. (2019)
OPF	Temperature = 173.4 °C Time = 3.3 h Catalyst = 10% Fe/HY zeolite	Levulinic acid	61.80	Ramli and Amin (2015)

(continued)

Table 2.2 (continued)

Feedstock	Experimental conditions	Product	Product yield (%)	References
OPF	Temperature = 30 °C Time = 96 h Microbe = *Kluyveromyces marxianus* ATCC 36907	Xylitol	35.00	Abdul Manaf et al. (2018)
OPF	Temperature = 75 °C Time = 70 min Catalyst = 70% Nitric acid	Oxalic acid	43.31	Maulina and Rahmadi (2017)

oil palm wastes into the chemical products are following the general steps as mentioned in Section (i). Bio-based chemicals such as succinic acid, lactic acid, xylitol, and 2,3-butanediol are fermentation products. Monomeric sugars derived from OPF and OPT are fed as nutrient substrates in an incubator for the growth of microbes. Meanwhile, bio-based chemicals such as furfural, levulinic acid, and oxalic acid are the products of chemical modification. In this case, monomeric sugars are used as reactive substrates in a reactor for the catalytic conversion.

Lactic acid synthesized using OPT sat as a nutrient source resulted in the highest yield of 92.00% among the bio-based chemicals (Kunasundari et al. 2017). This fermentation process could be operated at a relatively low temperature of 50 °C and 12 h. Time intensiveness was definitely one of the major drawbacks for the fermentation process. Aside from that, Ramli and Amin (2015) reported the levulinic acid production with a high yield of 61.80%. The given yield can be achieved through catalytic conversion at a high operating temperature of 173.4 °C and a relatively short reaction time of 3.3 h in the presence of 10% Fe/HY zeolite catalyst. This solid catalyst was claimed to be environmentally friendly and was able to be reused up to five times. Besides, oxalic acid was readily produced from OPF with a moderate yield of 43.31% through oxidation reaction (Maulina and Rahmadi 2017). However, oxalic acid is not ranked in the top platform chemicals as seen in Table 2.1. Even though oxalic acid might not be the desired chemical, it can be used to form choline chloride-oxalic acid, which is one type of deep eutectic solvent for lignocellulosic biomass utilization and conversion (Loow et al. 2017a). Recently, Lee et al. (2019) demostrated the potential application of aqueous choline chloride-oxalic acid in the production of furfural with a moderate yield up to 26.34% at 100 °C and 1 atm.

2.2.3 Biochar

Aside from the major trends of converting the OPF and OPT into bio-based energy and chemicals, there are a significant number of publications discussed on the production of biochar from these oil palm wastes. Biochar is the solid residue obtained when the biomass is heated at relatively low temperatures in an oxygen depleted environment. This thermochemical process of transforming the biomass into biochar is commonly termed as carbonization or pyrolysis. In addition, the pyrolysis process generates not only the solid product, but also fractions of liquid and gas called bio-oil and syngas (Alias et al. 2015). Bio-oil and syngas are potential energy products which can be served as alternatives to fossil fuel. This section highlights the properties and applications of biochar produced from two different oil palm wastes, OPF and OPT (Table 2.3).

(i) Properties of biochar

In general, biochar poses similar physiochemical properties as conventional charcoal such as richness in aromatic carbon, porosity, and high stability. Table 2.4 summarized the composition analysis of typical OPF, OPT and their respective biochars.

Table 2.3 Proximate and ultimate analysis of OPF, OPT, and their respective biochars (Kabir et al. 2017; Abdullah et al. 2016)

Properties	OPF	OPF biochar	OPT	OPT biochar
Proximate analysis (wt%)				
Moisture	4.83	4.42	64.30	6.66
Ash	5.87	3.73	3.57	12.96
Volatile matter	70.33	9.97	25.70	32.07
Fixed carbon	18.97	81.88	6.43	48.31
Heating value (MJ/kg)	16.00	24.15	4.60	21.00
Ultimate analysis (wt%)				
Carbon (C)	42.88	91.00	40.26	73.76
Hydrogen (H)	7.06	1.00	5.88	1.57
Nitrogen (N)	0.52	2.00	0.00	0.00
Oxygen (O)	49.54	6.00	53.86	24.67

Table 2.4 Brunauer-Emmett-Teller (BET) method analysis of oil palm waste biochars (Mahmood et al. 2015)

Properties	OPF biochar	Palm kernel shells (PKS) biochar	Empty fruit bunches (EFB) biochar
BET method analysis			
BET surface area (m^2/g)	857.31	23.73	95.83
Micropore area (m^2/g)	740.80	23.38	87.71
Micropore volume (cm^3/g)	0.35	0.01	0.04
Adsorption capacity (cm^3/g)	280.00	10.00	35.00

According to the guidelines of the European Biochar Certificate, solid residue generated from the pyrolysis process is qualified as biochar only if its carbon content is 50% and above (European Biochar Certificate 2012). Ultimate analysis that validated the carbon content in OPF and OPT was as high as 91.00% and 73.76%. This implied that both oil palm wastes fulfilled the prerequisite as they were able to form their very own biochars. Furthermore, the heating value of OPF and OPT biochars was determined to be 24.15 MJ/kg and 21.00 MJ/kg respectively as shown in Table 2.4. These values were comparable to that of conventional charcoal which is approximately 28.9 MJ/kg. Therefore, biochars derived from oil palm wastes were applicable as the potential solid biofuels. The significant reduction of volatile matter in OPF biochar indicates that most of the volatile matter in the original OPF was transferred into syngas. Interestingly, moisture content was the only composition that decreased for the case of OPT biochar. The presence of relatively high moisture content in the original OPT was probably the main cause of such an unusual trend.

(ii) Application of biochar

Biochar is applicable to a distinctive field of studies. One of the major applications of biochar is to mitigate environmental issues through sustainable approaches such as carbon sequestration and soil amendment (Lehmann et al. 2011; Manyà 2012). In fact, applications of biochar strongly associated with its properties. For instance, the rigid stability of biochar toward chemical and biological degradation makes it a suitable material for carbon sink (Manyà 2012). Meanwhile, the porous structure of biochar allows it to enhance soil productivity through improved water retention and soil aeration (Lehmann et al. 2011). Nevertheless, the properties of a given biochar depend on the characteristics of both process and feedstock applied during the production of biochar (Manyà 2012). This indicates that optimization on the characteristics of process and feedstock is required in order to produce biochar with desired properties that can meet the requirement for its applications. To date, the number of publications on the applications of biochar produced from OPF and OPT is limited. This is probably due to the difficulty in producing biochar with consistent properties which slows down the development.

Table 2.4 summarized the porous characteristics of biochar formed from different oil palm wastes. Mahmood et al. (2015) reported the porosity characteristics such as surface area, micropore area, micropore volume, and adsorption capacity of OPF biochar which were the highest when compared to the biochars produced from PKS and EFB. OPF biochar comprising of high porosity, large surface area, and exceptional adsorption capacity is, therefore, a potential candidate to be upgraded into high-quality activated carbon with wider applications. The applications of activated carbon involve multiple principles such as humidity regulator, supercapacitor, air purification, and wastewater treatment (Liew et al. 2018). The valorization of oil palm waste into valuable activated carbon is highly desired from the economic perspective. Maulina and Iriansyah (2018) reported the OPF biochar applied as an activated carbon was able to meet the quality standards of activated carbon according to the Indonesian National Standard, SNI 06-3730-1995.

In conclusion, biochar of similar properties is highly desired for the purpose of consistent performance. In addition, the properties of biochar used should be well reported throughout the research community to ensure a more adequate comparative study that could be conducted. Consequently, this will lead the researchers to an in-depth understanding of the relationship between biochar properties and its applications.

2.2.4 Fertilizer and Animal Feed

OPF and OPT have been attempted to be reused as fertilizer and/or animal feed in the past. For example, Loh et al. (2015) developed spent bleaching earth with OPF and OPT in different mixing ratios to form organic fertilizer. They found that the optimized blend of three components, namely spent bleaching earth, OPT, and chicken litter at the ratio of 1:1:0.5 exhibited good nutrient contents ($N:P_2O_5:K_2O =$ 0.65:1.59:1.63) and physicochemical properties (pH $= 5.4$, organic matter $= 40\%$, and C:N $= 36:1$) as a base material for organic fertilizer production. This blend could be further enhanced with mineral fertilizer to achieve a desired targeted NPK of 2:2:2 that encouraged the growth of vegetables (Loh et al. 2015). Masilamany et al. (2017) evaluated the phytotoxic effects of OPF mulch treated with imazethapyr at a reduced rate on weed emergence and growth. They found that imazethapyr was compatible with oil palm residue mulches. A field experiment in coconut plantation further revealed that imazethapyr-treated OPF mulch at a rate of 24 g a.i ha^{-1} + 3.4 t ha^{-1} provided excellent control of *Mikania micrantha*, *Asystasia gangetica*, *Phyllanthus amarus*, *Panicum* sp., and *Echinochloa colona* (Masilamany et al. 2017).

OPT was rarely used as an animal feed. Marlida et al. (2016) found that only the OPT which was treated with ligninase could be used as an animal feed, whereby the optimum concentration of OPT was 60% (v/w) and 750 units/kg of ligninase must be used to improve the fiber fractions for easy digestions of animals. Therefore, between OPF and OPT, OPF was more frequently to be tested as a potential source as an animal feed. Recently, Azmi et al. (2019) found that the isolated fungi from OPF showed enzyme activity profile that was suitable for the purpose as a pretreatment agent for the agricultural by-product to be used as an animal feed. Their results showed the isolated *Trichoderma harzianum* and *Fusarium solani* exhibited the optimal enzyme activity profile as a pretreatment agent compared to the white-rot fungi (Azmi et al. 2019). Hamchara et al. (2018) also found that fungus, namely *Lentinus sajor-caju*, could enhance the digestibility of OPF in goats. Hamchara et al. (2018) concluded that fungal treated OPF could be effectively used as an alternative roughage source in total mixed ration diets, constituting at least up to 100% of OPF. The addition of fungal-treated OPF also led to an increase in nitrogen retention (Hamchara et al. 2018). *Phanerochaete chrysosporium* was used to help pretreat OPF via the fermentation process before it was used as an animal feed (Febrina et al. 2017). According to Febrina et al. (2017), the pretreated OPF plus P, S, and Mg minerals could improve the consumption of nutrients and the growth performance of goats. The supplementation of P, S, and Mg might encourage the metabolic processes by rumen microbes that led to the utilization of nutrients in pretreated OPF (Febrina et al. 2017). Suryani et al. (2017) evaluated the effect of combining ammoniated OPF with direct-fed microorganisms, namely *Saccharomyces cerevisiae* and *Bacillus amyloliquefaciens*, and virgin coconut oil to produce a suitable animal feed for the Bali cattle. They found that the feed consisted of OPF which was supplemented with *S. cerevisiae* and virgin coconut oil increased the feed efficiency and reduced methane gas production by up to 20.63% as compared to the control (Suryani et al. 2017).

Generally, OPF could not be used as a single feed but must be combined with concentrates that had high protein and energy levels. For example, Jafari et al. (2018) found that OPF could be reused as a feed ingredient at 25% inclusion levels to support sheep farming, while the remaining concentrates mainly contained corn, soybean meal, palm kernel cake, rice bran, and other ingredients. The supplementation of 25% of OPF could result in improved rumen fermentation efficiency such as increased propionic acid concentration in the sheep (Jafari et al. 2018), while Ghani et al. (2017) found that the inclusion of OPF pellets in a complete animal feed increased the unsaturated fatty acid content in ruminants. Harahap et al. (2018) analyzed the quality of OPF using fiber cracking technology combined with *Indigofera* sp. in the ruminant ration to the fermentation in the rumen-simulating semicontinuous culture system. They found that the fiber cracking machine using urea for ammoniation could destroy the bonds of lignocellulose and lignohemicellulose, which were the factors causing low digestibility in ruminants (Harahap et al. 2018). Warly et al. (2017) determined the various levels of concentrate (consisted of rice bran, tofu waste, and ex-decanter solid waste from palm oil processing) and OPF on nutrient digestibility and apparent mineral bioavailability in beef cattle. They found that the diet containing 60% OPF resulted in a greater deficiency of minerals and all three experimental diets were deficient in minerals, showing that supplementation of certain minerals was needed to support the optimum production of beef cattle using OPF as an animal feed (Warly et al. 2017).

2.2.5 Other Products

Other applications of oil palm wastes involved the production of biopolymers. The most abundant biopolymer that can be extracted from OPT and OPF is termed as nanocellulose fiber or cellulose nanocrystal (Lamaming et al. 2015). In recent years, biopolymers derived from OPT and OPF showed great potential to be utilized as reinforcing agent, filler, and adhesive. In addition, a significant number of publications were dedicated to the fabrication of engineered wood products such as particleboard, composite board, and fiber board using OPT and OPF. Thus, this section is distinguished from the common applications of oil palm wastes as discussed in the sections above. The other applications of OPT and OPF in various fields are summarized in Table 2.5.

As seen in Table 2.5, OPF is able to act as a good reinforcing fiber for the poly (vinyl alcohol) (PVA) based composite (Sukudom et al. 2019). On the other hand, gypsum filled with 20% OPT loading showed great improvement in comparison

Table 2.5 Other applications of OPT and OPF

Feedstock	Application	Results	References
OPF	Reinforcing agent	Improved mechanical, thermal, and water-resistance properties of the composite with the presence of OPF reinforcing agent	Sukudom et al. (2019)
OPT	Filler	Improved mechanical, thermal, and fire-retardant properties of the gypsum with the presence of 20% OPT filler loading	Selamat et al. (2019)
OPT	Adhesive	Improved adhesion strength of the adhesive produced from OPT starch with the presence of nanozeolite	Dur et al. (2018)
OPF	Fibreboard	Improved sound-absorbing properties at different frequencies of the fibreboard produced from OPF with an optimum density of 0.17 g cm^{-3}	Bubparenu et al. (2018)
OPT	Particleboard	Improved thermal and fire-retardant properties of the particleboard produced from OPT with the presence of 20% magnesium oxide	Selamat et al. (2018)
OPT	Composite board	Improved mechanical, physical, and morphology properties of composite board produced from OPT at the presence of a 3% sodium hydroxide solution	Nasution et al. (2018)

to gypsum without OPT filler (Selamat et al. 2019). However, gypsum produced with OPT filler poses hygroscopic properties which might have an adverse effect on its durability when in contact with water or under a high moisture environment. Besides nanocellulose, starch is yet another type of biopolymer that can be found in OPT. Dur et al. (2018) demonstrated the fabrication of adhesive using OPT starch filled with nanozeolite. Nanozeolite played an important role to fully enclose the surface and thus no cavities appeared on the adhesive. Finally, applications of both biopolymers and engineered wood products are definitely another potential route for the valorization of oil palm wastes.

2.3 Conclusion

The conversion of biomass into various value-added biochemicals is generally termed as biomass valorization. As the demand for human consumption, agriculture, and industries grows, the most easily reachable resources become insufficient. Thus, more and more research on the valorization of agricultural residual into biofuel and bio-based products has attracted significant attention of late, mainly due to the positive effects from both economic and environmental aspects and long-term energy sustainability with greenhouse gas mitigation. In Malaysia, the huge production of crude palm oil is leading to a huge amount of oil palm agriculture waste being produced. OPF and OPT are two by-products produced from the oil palm plantation, which are also the highest lignocellulosic biomass generated in Malaysia. The reuse and valorization of OPF and OPT into bioenergy, bio-based chemicals, biochar, fertilizer, animal feed, and other products are seen as sustainable ways to manage those wastes. Although the emphasis on the sustainability of OPF and OPT reuse is well intentioned, it may sometimes raise false expectations. The limits imposed by the economic and social frameworks may present limitations to sustainable practices that could be applied to the large scale. Biomass reuse and valorization are therefore limited unless the activity is subsidized, reusability of OPF and OPT is successfully designed for commercial reuse and, most importantly, the government takes the initiative in legislating for sustainable industrial development. The initiatives to promote OPF and OPT reuse can only come from the authorities, institutions, and decision-makers since their subtle actions could accelerate the research and development needed for promoting sustainability in oil palm plantations.

References

Abdul Manaf, S. F., Md Jahim, J., Harun, S., & Luthfi, A. A. I. (2018). Fractionation of oil palm fronds (OPF) hemicellulose using dilute nitric acid for fermentative production of xylitol. *Industrial Crops and Products, 115,* 6–15.

Abdullah, H., Jie, W. S., Yusof, N., & Isa, I. M. (2016). Fuel and ash properties of biochar produced from microwave-assisted carbonisation of oil palm trunk core. *Journal of Oil Palm Research, 28,* 81–92.

Ahmad, F. B., Zhang, Z., Doherty, W. O. S., & O'Hara, I. M. (2019). The prospect of microbial oil production and applications from oil palm biomass. *Biochemical Engineering Journal,* 9–23.

Alias, N. B., Ibrahim, N., Hamid, M. K. A., Hasbullah, H., Ali, R. R., & Kasmani, R. M. (2015). Investigation of oil palm wastes' pyrolysis by thermo-gravimetric analyzer for potential biofuel production. *Energy Procedia,* 78–83.

Ang, S. K., Adibah, Y., Abd-Aziz, S., & Madihah, M. S. (2015). Potential uses of xylanase-rich lignocellulolytic enzymes cocktail for oil palm trunk (OPT) degradation and lignocellulosic ethanol production. *Energy & Fuels, 29,* 5103–5116.

Azmi, M. A., Yusof, M. T., Zunita, Z., & Hassim, H. A. (2019). Enhancing the utilization of oil palm fronds as livestock feed using biological pre-treatment method. In *IOP Conference Series: Earth and Environmental Science.*

Bardant, T. B., Winarni, I., & Sukmana, H. (2017). High-loading-substrate enzymatic hydrolysis of palm plantation waste followed by unsterilized-mixed-culture fermentation for bio-ethanol production. In *AIP Conference Proceedings.*

Brandt, A., Gräsvik, J., Hallett, J. P., & Welton, T. (2013). Deconstruction of lignocellulosic biomass with ionic liquids. *Green Chemistry, 15,* 550–583.

Bubparenu, N., Laemsak, N., Chitaree, R., & Sihabut, T. (2018). Effect of density and surface finishing on sound absorption of oil palm frond. *Asia-Pacific Journal of Science and Technology, 23.*

Bukhari, N. A., Jahim, J. M., Loh, S. K., Bakar, N. A., & Luthfi, A. A. I. (2019). Response surface optimisation of enzymatically hydrolysed and dilute acid pretreated oil palm trunk bagasse for succinic acid production. *BioResources, 14,* 1679–1693.

Chandel, A. K., Garlapati, V. K., Singh, A. K., Antunes, F. A. F., & da Silva, S. S. (2018). The path forward for lignocellulose biorefineries: Bottlenecks, solutions, and perspective on commercialization. *Bioresource Technology, 264,* 370–381.

Cherubini, F. (2010). The biorefinery concept: Using biomass instead of oil for producing energy and chemicals. *Energy Conversion and Management, 51,* 1412–1421.

Dur, S., Daulay, A. H., Padli Nasution, M. I., Sari, R. F., & Furqan, M. (2018). Preparation and properties nanozeolite-filled modified oil palm trunk starch nanocomposites. In *Journal of Physics: Conference Series.*

EBC. (2012). *European biochar certificate—Guidelines for a sustainable production of biochar.* Arbaz, Switzerland: European Biochar Foundation (EBC). http://www.europeanbiochar.org/en/download. Version 8.1E of 4th April 2019, https://doi.org/10.13140/rg.2.1.4658.7043.

Eom, I. Y., Oh, Y. H., Park, S. J., Lee, S. H., & Yu, J. H. (2015a). Fermentative l-lactic acid production from pretreated whole slurry of oil palm trunk treated by hydrothermolysis and subsequent enzymatic hydrolysis. *Bioresource Technology, 185,* 143–149.

Eom, I. Y., Yu, J. H., Jung, C. D., & Hong, K. S. (2015b). Efficient ethanol production from dried oil palm trunk treated by hydrothermolysis and subsequent enzymatic hydrolysis. *Biotechnology for Biofuels, 8.*

Farah Amani, A. H., Toh, S. M., Tan, J. S., & Lee, C. K. (2018). The efficiency of using oil palm frond hydrolysate from enzymatic hydrolysis in bioethanol production. *Waste and Biomass Valorization, 9,* 539–548.

Febrina, D., Jamarun, N., Zain, M., & Khasrad. (2017). Effects of using different levels of oil palm fronds (Fopfs) fermented with phanerochaete chrysosporium plus minerals (p, s and mg) instead of napier grass on nutrient consumption and the growth performance of goats. *Pakistan Journal of Nutrition, 16,* 612–617.

Ghani, A. A. A., Rusli, N. D., Shahudin, M. S., Goh, Y. M., Zamri-Saad, M., Hafandi, A., et al. (2017). Utilisation of oil palm fronds as ruminant feed and its effect on fatty acid metabolism. *Pertanika Journal of Tropical Agricultural Science, 40,* 215–224.

Hamchara, P., Chanjula, P., Cherdthong, A., & Wanapat, M. (2018). Digestibility, ruminal fermentation, and nitrogen balance with various feeding levels of oil palm fronds treated with Lentinus sajor-caju in goats. *Asian-Australasian Journal of Animal Sciences, 31,* 1619–1626.

Harahap, R. P., Jayanegara, A., Nahrowi, & Fakhri, S. (2018). Evaluation of oil palm fronds using fiber cracking technology combined with Indigofera sp. in ruminant ration by Rusitec. In *AIP Conference Proceedings.*

Hazeena, S. H., Pandey, A., & Binod, P. (2016). Evaluation of oil palm front hydrolysate as a novel substrate for 2,3-butanediol production using a novel isolate Enterobacter cloacae SG1. *Renewable Energy, 98,* 216–220.

Ho, M. C., Ong, V. Z., & Wu, T. Y. (2019). Potential use of alkaline hydrogen peroxide in lignocellulosic biomass pretreatment and valorization–A review. *Renewable and Sustainable Energy Reviews, 112,* 75–86.

Ho, M. C., & Wu, T. Y. (2020). Sequential pretreatment with alkaline hydrogen peroxide and choline chloride:copper (II) chloride dihydrate–Synergistic fractionation of oil palm fronds. *Bioresource Technology, 301,* 122684.

Jafari, S., Meng, G. Y., Rajion, M. A., Torshizi, M. A. K., & Ebrahimi, M. (2018). Effect of supplementation of oil palm (Eleis guineensis) frond as a substitute for concentrate feed on rumen fermentation, carcass characteristics and microbial populations in sheep. *Thai Journal of Veterinary Medicine, 48,* 9–18.

Kabir, G., Mohd Din, A. T., & Hameed, B. H. (2017). Pyrolysis of oil palm mesocarp fiber and palm frond in a slow-heating fixed-bed reactor: A comparative study. *Bioresource Technology, 241,* 563–572.

Kumar, D., Singh, B., & Korstad, J. (2017). Utilization of lignocellulosic biomass by oleaginous yeast and bacteria for production of biodiesel and renewable diesel. *Renewable and Sustainable Energy Reviews, 73,* 654–671.

Kunasundari, B., Arai, T., Sudesh, K., Hashim, R., Sulaiman, O., Stalin, N. J., et al. (2017). Detox-ification of sap from felled oil palm trunks for the efficient production of lactic acid. *Applied Biochemistry and Biotechnology, 183,* 412–425.

Lai, L. W., & Idris, A. (2013). Disruption of oil palm trunks and fronds by microwave-alkali pretreatment. *BioResources, 8*(2), 2792–2804.

Lamaming, J., Hashim, R., Sulaiman, O., Leh, C. P., Sugimoto, T., & Nordin, N. A. (2015). Cellulose nanocrystals isolated from oil palm trunk. *Carbohydrate Polymers, 127,* 202–208.

Lee, C. B. T. L., Wu, T. Y., Ting, C. H., Tan, J. K., Siow, L. F., Cheng, C. K., Md. Jahim, J., & Mohammad, A. W. (2019). One-pot furfural production using choline chloride-dicarboxylic acid based deep eutectic solvents under mild conditions. *Bioresource Technology, 278,* 486–489.

Lehmann, J., Rillig, M. C., Thies, J., Masiello, C. A., Hockaday, W. C., & Crowley, D. (2011). Biochar effects on soil biota—A review. *Soil Biology & Biochemistry, 43,* 1812–1836.

Liew, R. K., Nam, W. L., Chong, M. Y., Phang, X. Y., Su, M. H., Yek, P. N. Y., et al. (2018). Oil palm waste: An abundant and promising feedstock for microwave pyrolysis conversion into good quality biochar with potential multi-applications. *Process Safety and Environmental Protection, 115,* 57–69.

Loh, S. K., Cheong, K. Y., Choo, Y. M., & Salimon, J. (2015). Formulation and optimisation of spent bleaching earth-based bio organic fertiliser. *Journal of Oil Palm Research, 27,* 57–66.

Loh, S. K. (2017). The potential of the Malaysian oil palm biomass as a renewable energy source. *Energy Conversion and Management, 141,* 285–298.

Loow, Y. L., New, E. K., Yang, G. H., Ang, L. Y., Foo, L. Y. W., & Wu, T. Y. (2017a) Potential use of deep eutectic solvents to facilitate lignocellulosic biomass utilization and conversion. *Cellulose, 24*(9), 3591–3618.

Loow., Y. L., & Wu, T. Y., Lim, Y. S., Tan, K. A., Siow, L. F., Jahim, J. M., & Mohammad, A. W. (2017b). Improvement of xylose recovery from the stalks of oil palm fronds using inorganic salt and oxidative agent. *Energy Conversion and Management, 138,* 248–260.

Loow, Y. L., Wu, T. Y., Yang, G. H., Ang, L. Y., New, E. K., Siow, L. F., Jahim, J. M., Mohammad, A. W., & Teoh, W. H. (2018). Deep eutectic solvent and inorganic salt pretreatment of lignocellulosic biomass for improving xylose recovery. *Bioresource Technology, 249,* 818–825.

Luthfi, A. A. I., Tan, J. P., Harun, S., Manaf, S. F. A., & Jahim, J. M. (2019). Homogeneous solid dispersion (HSD) system for rapid and stable production of succinic acid from lignocellulosic hydrolysate. *Bioprocess and Biosystems Engineering, 42,* 117–130.

Mahmood, W. M. F. W., Ariffin, M. A., Harun, Z., Ishak, N. A. I. M., Ghani, J. A., & Rahman, M. N. A. (2015). Characterisation and potential use of biochar from gasified oil palm wastes. *Journal of Engineering Science and Technology, 10,* 45–54.

Manyà, J. J. (2012). Pyrolysis for biochar purposes: A review to establish current knowledge gaps and research needs. *Environmental Science and Technology, 46,* 7939–7954.

Marlida, Y., Arnim, & Roza, E. (2016). The effect treated of oil palm trunk by ligninase thermostable to improvement fiber quality as energy sources by ruminant. *International Journal of ChemTech Research, 9,* 429–436.

Masilamany, D., Mat, M. C., & Seng, C. T. (2017). The potential use of oil palm frond mulch treated with imazethapyr for weed control in Malaysian coconut plantation. *Sains Malaysiana, 46,* 1171–1181.

Mastuli, M. S., Kamarulzaman, N., Kasim, M. F., Sivasangar, S., Saiman, M. I., & Taufiq-Yap, Y. H. (2017). Catalytic gasification of oil palm frond biomass in supercritical water using MgO supported Ni, Cu and Zn oxides as catalysts for hydrogen production. *International Journal of Hydrogen Energy, 42*, 11215–11228.

Maulina, S., & Iriansyah, M. (2018). Characteristics of activated carbon resulted from pyrolysis of the oil palm fronds powder. In *IOP Conference Series: Materials Science and Engineering*.

Maulina, S., & Rahmadi, I. (2017). The utilization of oil palm fronds in producing oxalic acid through oxidation. In *AIP Conference Proceedings*.

Mohd Zakria, R., Gimbun, J., Asras, M. F. F., & Chua, G. K. (2017). Magnesium sulphate and B-alanine enhanced the ability of Kluyveromyces marxianus producing bioethanol using oil palm trunk sap. *Biofuels, 8*, 595–603.

Nasution, D. Y., Marpongahtun, Gea, S., Ardiansyah, & Ridho. (2018). Characterization of composite boards made of oil palm trunk flour/maleic anhydride grafted polypropylene. In *Journal of Physics: Conference Series*.

Ong, V. Z., Wu, T. Y., Lee, C. B. T. L., Cheong, N. W. R., & Shak, K. P. Y. (2019). Sequential ultrasonication and deep eutectic solvent pretreatment to remove lignin and recover xylose from oil palm fronds. *Ultrasonics Sonochemistry, 58*, 104598.

Ramli, N. A. S., & Amin, N. A. S. (2015). Optimization of renewable levulinic acid production from glucose conversion catalyzed by Fe/HY zeolite catalyst in aqueous medium. *Energy Conversion and Management, 95*, 10–19.

Ramli, N. A. S., & Amin, N. A. S. (2017). Optimization of biomass conversion to levulinic acid in acidic ionic liquid and upgrading of levulinic acid to ethyl levulinate. *Bioenergy Research, 10*, 50–63.

Selamat, M. E., Hashim, R., Sulaiman, O., Kassim, M. H. M., Saharudin, N. I., & Taiwo, O. F. A. (2019). Comparative study of oil palm trunk and rice husk as fillers in gypsum composite for building material. *Construction and Building Materials, 197*, 526–532.

Selamat, M. E., Hui, T. Y., Hashim, R., Sulaiman, O., Kassim, M. H. M., & Stalin, N. J. (2018). Properties of particleboard made from oil palm trunks added magnesium oxide as fire retardant. *Journal of Physical Science, 29*, 59–75.

Sitthikitpanya, S., Reungsang, A., & Prasertsan, P. (2018). Two-stage thermophilic bio-hydrogen and methane production from lime-pretreated oil palm trunk by simultaneous saccharification and fermentation. *International Journal of Hydrogen Energy, 43*, 4284–4293.

Sitthikitpanya, S., Reungsang, A., Prasertsan, P., & Khanal, S. K. (2017). Two-stage thermophilic bio-hydrogen and methane production from oil palm trunk hydrolysate using Thermoanaerobacterium thermosaccharolyticum KKU19. *International Journal of Hydrogen Energy, 42*, 28222–28232.

Sukudom, N., Jariyasakoolroj, P., Jarupan, L., & Tansin, K. (2019). Mechanical, thermal, and biodegradation behaviors of poly(vinyl alcohol) biocomposite with reinforcement of oil palm frond fiber. *Journal of Material Cycles and Waste Management, 21*, 125–133.

Suryani, H., Zain, M., Ningrat, R. W. S., & Jamarun, N. (2017). Effect of dietary supplementation based on an ammoniated palm frond with direct fed microbials and virgin coconut oil on the growth performance and methane production of bali cattle. *Pakistan Journal of Nutrition, 16*, 599–604.

Warly, L., Suyitman, Evitayani, & Fariani, A. (2017). Nutrient digestibility and apparent bioavailability of minerals in beef cattle fed with different levels of concentrate and oil-palm fronds. *Pakistan Journal of Nutrition, 16*, 131–135.

Chapter 3
Palm Oil Milling Wastes

3.1 Introduction

This chapter focuses on the wastes that are generated specifically from palm oil milling activities. We will first introduce the palm oil milling processes and also identify processes which generate wastes. Subsequently, we will look into how wastes from palm oil milling are currently managed, any governing laws and limitations to how these wastes could be handled. Finally, a comprehensive overview on technological advancements on palm oil mill wastes and future directions of palm oil mill waste management will be discussed here.

3.2 Palm Oil Milling Process

Extraction of crude palm oil (CPO) can be conducted in two different modes, namely dry and wet modes, where the latter is more commonly employed for palm oil milling in Malaysia (Wu et al. 2010). A schematic of a wet palm oil milling process is illustrated in Fig. 3.1, with descriptions of individual unit operations of the process as follows:

3.3 Sterilization

This is the first step upon receiving the fresh fruit bunches delivered from plantations. The aim of sterilization is to (i) deactivate enzymes that will contribute in the rapid formation of free fatty acids, leading to better CPO yield from the process; (ii) and loosen fruits that are still attached to fresh fruit bunches. This process uses steam which lasts for 50 min at 140 °C and 30 kPa in horizontal sterilizers (Wu et al. 2010; Liew et al. 2015). This process was also found to smoothen the process of stripping to

© Springer Nature Switzerland AG 2020
P. E. Poh et al., *Waste Management in the Palm Oil Industry*,
Green Energy and Technology, https://doi.org/10.1007/978-3-030-39550-6_3

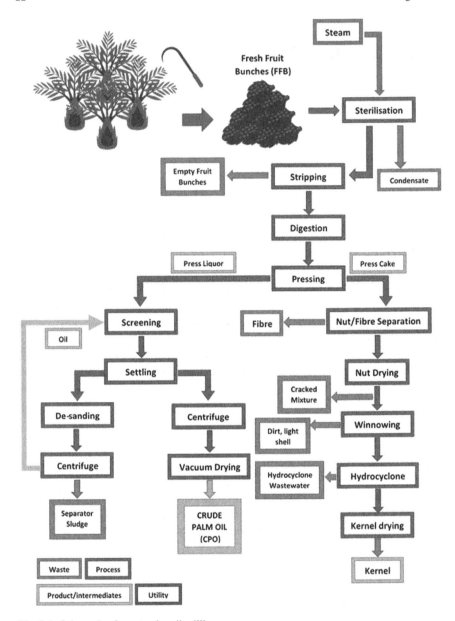

Fig. 3.1 Schematic of a wet palm oil milling process

free the fruits from bunches. In this process, condensate from the sterilizer contributes to the wastewater stream generated from the palm oil milling process.

3.4 Stripping, Digestion, and Pressing

Fruit bunches exiting the sterilizer then enters a rotary drum thresher to have the fruits detached from the bunch stalks. This process generates solid wastes in the form of empty fruit bunches which will be discussed in later sections of this chapter. Meanwhile, the sterilized fruits that were separated from the bunch proceed to be digested in a vessel at 80–90 °C, where a middle rotating shaft works to further soften the fruits and loosen the mesocarp from the nuts (Ahmad et al. 2016). After digestion, the fruits are then sent to a screw press to extract palm oil.

3.5 Clarification

Palm oil that exits the screw press is regarded as press liquor. The press liquor contains impurities such as water, debris from fruits, and fibrous materials. Press liquor is sent to a series of processes to obtain CPO that will be used in subsequent oleochemical processes—this is regarded as clarification.

Firstly, the press liquor undergoes screening to remove large debris, followed by settling to separate the oil and water layer. The top layer in the settling vessel would be the oil that has lower density, while the bottom layer consists of water and non-dissolved solids. The bottom layer is sent to a centrifuge to recover additional oil that contributes to improving the oil extraction rate in the milling process. The bottom layer of the centrifuge then contributes to the production of another stream of liquid waste that is eventually combined with the condensate from sterilization and hydrocyclone wastewater to form palm oil mill effluent (POME).

The top layer from settling vessel which consists mostly of CPO is sent to a centrifuge to further purification of CPO. Subsequently, the layer of CPO that is extracted from the centrifuge then undergoes vacuum drying at mild temperatures to remove excess moisture, producing an end product (CPO) that is being sent to oleochemical plants as raw materials for other processes.

3.6 Kernel Extraction Process

The screw press also produces a cake which constitutes of nuts and fiber. This cake is referred to as the press cake in Fig. 3.1. The press cake undergoes separation of nuts from the fiber mechanically or with the use of an air stream, leading to the generation of fibrous wastes.

Next, the separated nuts undergo cracking, drying, and winnowing where waste of cracked shells is generated. Subsequently, the nuts are put through hydrocyclone for separation of empty shells from kernels. The hydrocyclone unit utilizes water which also contributes to the production of palm oil mill effluent from the milling process. Finally, the kernel is dried before further processing in a separate process.

3.7 Solid Waste Management in Palm Oil Mills

A typical of fresh fruit bunch (FFB) contains approximately 21% crude palm oil (CPO) and the remaining are the empty fruit bunch (22–23%), mesocarp fiber (13.5 15%), palm kernel shell (5.5–7%), and palm kernel (6–7%) (Kong et al. 2014). A staggering amount of solid biomass waste which consists mainly of empty fruit bunches (EFB), palm pressed fiber (PPF), and palm kernel shell (PKS) is generated after CPO extraction. For every ton of CPO extracted, approximately 1.8 tons of the solid biomass waste is produced (Irvan 2018). Based on the capacity of a typical palm oil mill, an estimated 4–24 tons of solid biomass waste would be generated on an hourly basis (Mahlia et al. 2001).

The major source of the solid biomass waste is from EFB, which is the by-product produced in the stripping process during palm oil processing. Individual fruitlets from FFB are stripped off from the brunches in a rotating drum and sent to the digester, leaving the EFB. PPF is another solid biomass waste produced from the palm oil mill. PPF is produced as a result of the removal of nuts from the fiber. PPF contains mainly of mesocarp fiber, some kernel shell fragments, and broken kernels. Generally, PPF contains about 5–6% of residue oil after extraction of CPO (Lau et al. 2006). Besides, PKS, the shell fractions remain, is produced as a result of nut cracking to recover the palm kernel. Figure 3.2 is a summary of solid waste management strategies in palm oil mill, which will be further discussed below.

3.7.1 Incineration

The traditional means to dispose EFB is by burning the empty bunches and return the ashes as fertilizer. The ash produced contains approximately 30–40% potassium oxide (K_2O) and has been shown to improve the oil palm plantation on acidic coastal soils (Singh et al. 2010). However, due to the air pollution associated with this practice, the open burning of EFB, without energy recovery, has been discouraged and banned in countries like Malaysia and Indonesia. Apart from that, EFB can be used as fuel in low-grade boilers in the mills to provide steam for the sterilization process. EFB by nature is a poor fuel due to its high moisture content (approx. 60%) and they are typically air-dried in order to reduce 40% of its original moisture content for more efficient combustion. Moreover, the utilization of EFB for power generation requires preprocessing such as shredding and pressing, which incur cost (Samiran

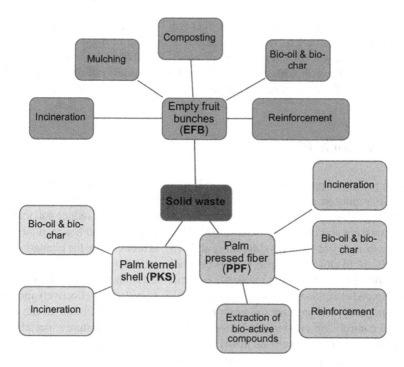

Fig. 3.2 Summary of solid waste management in palm oil mill

et al. 2016). Instead, most plants prefer burning of EFB to produce ash. The burnt ash is then returned as fertilizer in plantation due to its high potassium content. A major environmental concern on the burning of EFB is that it can result in the emission of "white smoke" to the environment owing to the significant quantity of moisture in EFB.

It is a common practice for PPF to be mixed with PKS and used as a solid fuel for heat and electricity generation in the mill. PPF and PKS, which contains an appreciable amount of lignin, are good sources of fuel for incineration. From Table 3.1, it is evident that PPF and PKS have relatively higher calorific value as compared to EFB. In Malaysia, the mills have been combusting 70% PPF and 30% PKS in their boilers to produce steam and generate electricity to satisfy the power consumption

Biomass	Heating value (MJ/kg) (dry weight basis)	Moisture content (%)
Empty fruit bunches	18.88	60–67
Palm pressed fiber	19.06	30–37
Palm kernel shell	20.09	12–20

Table 3.1 Heating values of solid biomass waste (Kong et al. 2014; Sabil et al. 2013)

in the mill (Sukiran et al. 2017). Excess fiber will be transported together with EFB to the plantation as mulch. The main concern of burning of these biomasses is the emission of dark smoke as a result of carryover of carbonized fibrous particulates from the boiler. Nevertheless, the incineration of solid wastes (PPF and PKS) in palm oil mills allows the mills to be self-sustainable in terms of energy usage.

3.7.2 Return to Plantation

A more practical approach is to return EFB in oil palm plantation due to the large quantities produced. EFB can be used as organic mulch in the plantation. Mulching is the practice of applying a layer of materials on the soil surface to reduce soil temperature, conserve soil moisture, increase soil pH and nutrients, and hence, improving the growth and yield of plants (Moradi et al. 2015). In areas where the plantation and the mills are in close proximity, EFB is generally used as mulch in the plantations due to convenience. However, incineration is still the common practice in mills with no associated plantation. The practice of using EFB as mulch can effectively increase soil aggregation and improve soil water retention. In addition, EFB mulch can assist in weed control, soil erosion prevention, and maintain soil moisture especially for young oil palm (Awalludin et al. 2015). EFB contains essential plant nutrients such as nitrogen (N), phosphorus (P), potassium (K), magnesium (Mg), calcium (Ca), and carbon (C) (Abu Bakar et al. 2011). The soil properties can be improved when these elements are released into the soil during the decomposition of the mulch. Nonetheless, mulching is usually supplemented with inorganic fertilizers due to the slow release of nutrients from the EFB (Lim and Rahman 2002). One major disadvantage of EFB is its bulkiness, rendering the storage, transportation, and practical application rather inconvenient and expensive.

EFB can be composted to reduce its volume and facilitate its application in the plantation, hence reducing the cost. Composted EFB could have its initial volume and weight reduced by up to 85% (Salètes et al. 2004). Additionally, composted EFB could increase soil C:N ratio and contain micronutrients that could be beneficial when returned to plantation as fertilizer (Siddiquee et al. 2017). It has also been reported that composting of a combination of EFB, palm oil mill effluent (POME) (Mohammad et al. 2012), decanter cake (Yahya et al. 2010), and animal feces (chicken manure, cow dung, goat dung) (Thambirajah et al. 1995) can improve the quality of the compost. It is worth noting that composting is a biological process usually requires a duration of 2–3 months to complete the composting process. Moreover, the proper control of the microbial activity is required to produce high-quality compost.

3.7.3 Conversion to Value-Added Products

Solid biomass wastes (EFB, PPF, PKS) generated in the mill are lignocellulosic in nature. They contain a high amount of cellulose, hemicellulose, and lignin (Chan et al. 2014). As with other lignocellulosic biomass, it can be transformed into a variety of useful compounds for industrial applications. Various technologies have been employed to convert solid biomass waste in palm oil mill into value-added products. Solid biomass wastes (EFB, PPF, PKS) can be converted into bio-oil and biochar through pyrolysis. Pyrolysis is a thermal decomposition of lignocellulosic biomass, such as oil palm biomass, into gaseous and liquid fuels in the absence of oxygen (Onoja et al. 2018). The process usually takes place at a temperature of about 400–600 °C. The products of the pyrolysis of oil palm biomass are condensable organic liquids (bio-oil), non-condensable gasses (carbon monoxide, carbon dioxide, hydrogen, methane), and biochar. The pyrolysis product is highly dependent on the process conditions and the cellulose, hemicellulose, and lignin content in the biomass. In general, there are two types of pyrolysis, namely fast pyrolysis and slow pyrolysis. Fast pyrolysis mainly yields bio-oil (70%), followed by biochar (15%), and non-condensable gasses (13%) (Kong et al. 2014). Among the solid biomass waste produced in the mill, PKS yields the highest amount of bio-oil using as compared to EFB and PPF due to the significantly higher amount of lignin present in PKS (Abnisa et al. 2013; Chan et al. 2014). If the biomass is decomposed under slow pyrolysis, mostly biochar would be produced instead.

Bio-oil is a dark brownish, free-flowing, oxygenated liquid that comprises molecules of various sizes as a result of the depolymerization and fragmentation reactions of lignin, hemicellulose, and cellulose. The bio-oil produced from EFB mainly comprises phenol, furan, ketone, alcohol, acids, and pyrans while the major components of bio-oil produced from PKS are phenols, acids, ketones, alcohols, and furans (Dhyani and Bhaskar 2018). Bio-oil can be used as fuel in engines, gas turbine, and boilers after upgradation. Besides, bio-oil can be used as feedstock to produce many specialty chemicals due to the presence of functional groups such as carbonyl, carboxyl, and phenolic groups. The challenge to substitute bio-oil as petroleum and chemical feedstock remains. Bio-oil from oil palm biomass contains an enormous amount of moisture (40–70%), depending on whether EFB, PKS, or PPF is used as the biomass feedstock (Abnisa et al. 2013). Besides, the decomposition of lignin and polysaccharides in the biomass result in a high quantity of unfavorable oxygenated compounds, such as carboxylic acids, ketones, and aldehydes, in the bio-oil (Dhyani and Bhaskar 2018). Such compounds contribute to properties such as corrosiveness, instability, immiscibility with conventional and low heating value of bio-oil. Therefore, proper upgradation of bio-oil, such as hydrodeoxygenation, is still required to ensure widespread application of the bio-oil.

While the main product of palm biomass waste under fast pyrolysis is bio-oil, those that of slow pyrolysis, however, are mostly biochar. Biochar is the carbon-rich residue remaining after the pyrolysis process. The biochars produced from biomass waste (such as EFB, PPF, PKS) have high heating values (20–30 MJ/kg), which is comparable to those of commercially available coal (Abnisa et al. 2013). Besides, biochar can be added to the soil to improve soil fertility. The porous structure of biochar helps to promote nutrient retention, improve soil pH and water holding capacity. The biochar in the soil can slowly release adsorbed moisture and nutrients to the soil due to the high adsorption capacity. Apart from that, biochar produced from solid biomass waste can act as a carbon sequester. Carbon dioxide (CO_2) from the atmosphere is captured by plants such as oil palm trees for photosynthesis and released during biomass combustion. After biomass combustion, only 3% of the initial carbon is retained while the rest is returned to the atmosphere. The conversion of biomass to biochar, however, retains 50% of the initial carbon (Lehmann et al. 2006). Moreover, biochar is highly stable in soil and allows the carbon to be stored for many years. Alternatively, biochar can be used as adsorbents for the removal of water pollutants such as toxic heavy metals, synthetic dyes, and pharmaceuticals (Rangabhashiyam and Balasubramaniam 2019). Biochar can be activated by physical means such as gas or steam activation to improve properties such as specific surface area, pore volume, and pore structures. Activated PKS derived biochar demonstrated enhanced adsorption capacity as compared to nonactivated biochar (Nasri et al. 2014).

Alternatively, solid biomass waste such as EFB, PPF, and PKS can be converted into briquettes to increase its energy density. Briquetting is a process of compacting loose oil palm biomass into a homogenous and densified solid. The biomass can be densified into briquettes by applying pressure via screw extrusion technique at high temperature and pressure with or without binder (Sulaiman et al. 2011). Pulverized EFB briquettes have exhibited good combustion characteristics comparable to sawdust briquettes and terrified wood (Nasrin et al. 2008). EFB can be blended with sawdust to produce better briquettes. Briquettes produced from PPF and PKS also exhibited good mechanical properties as a fuel (Husain et al. 2002). The advantages of briquetting include an increase in energy density per unit volume, ease and reduce the cost of handling and transportation. The briquettes can be used for industrial heating unit operations due to simplified handling and transportation. Nevertheless, it is difficult for the briquettes to compete in the market due to the availability of low-cost charcoal and wood, especially in a rural setting.

Apart from pyrolysis, gasification is one of the techniques that transform solid biomass waste into value-added products. Gasification is the conversion of biomass into various gaseous compounds via partial oxidation of the biomass (Onoja et al. 2018). The reaction usually occurs at high temperatures (800–1800 °C) (Awalludin et al. 2015). The main products from biomass gasification are hydrogen (H_2), carbon monoxide (CO), methane (CH_4), carbon dioxide (CO_2), and nitrogen (N_2). This mixture of gaseous products, sometimes known as syngas, can be used directly as fuel or further converted into liquid hydrocarbon via Fischer–Tropsch process. EFB and PKS can be used as feedstock for gasification. Hydrolysis of hemicellulose in the biomass favors the formation of CO_2 and H_2 while cellulose and lignin favors the formation of CO and CH_4 (Samiran et al. 2016). However, syngas from biomass also contains a substantial quantity of impurities such as acidic gasses, alkali compounds, and tar. Further, gas cleaning process is required to reduce the amount of impurities for commercial use.

Recently, the liquefaction process has gained an appreciable amount of attention due to its ability to produce valuable compounds such as bio-oil and resin precursor. Liquefaction is the breaking down of cellulose, hemicellulose, and lignin present in lignocellulosic biomass into reactive low molecular weight compounds useful in the industry. Liquefaction can be performed at a temperature of 120–250 °C in organic solvents with or without the presence of a catalyst (base or acid). The liquefaction of EFB and PPF to bio-oil has been studied using different alkaline catalysts and different solvents. However, the research of liquefaction of palm biomass is still at its infancy and more work needs to be done before it can be commercially viable.

3.7.4   Other Applications

There are multiples instances where solid biomass waste, such as EFB, be processed into fibrous form is used as a precursor for various applications. One of these applications is the production of pulp and paper. Studies have shown promising results in using EFB to produce pulp and paper. For instance, the quality of the paper obtained from pulps of EFB is comparable to hardwood Kraft pulp (Singh et al. 2013). The soda pulp produced from EFB is suitable to produce papers and corrugated cartons (Wan Daud and Law 2010). A pilot-scale production has also demonstrated that high-quality soda–anthraquinone pulp can be produced from EFB (Sharma et al. 2015). In another example, EFB and PPF are used as a filler to reinforce polymer composites using melt blending and hot-press molding techniques. The results show that PPF is a better reinforcing agent as compared to EFB (Then et al. 2013). EFB can also be processed in fibrous material and use in fiber board manufacturing. EFB is used to produce medium density fiber (MDF). Other products of EFB are coir fiber, fiber board, cement board, roofing tile, and card paper (Prasertsan and Prasertsan 1996). Recently, researchers have also demonstrated the isolation of cellulose nanofiber, microcrystalline cellulose, and nanocrystalline cellulose from EFB via acid hydrolysis (Fahma et al. 2010; Hassan et al. 2019; Hastuti et al. 2018). Cellulose has a wide range of applications in various industries such as foods stabilizer, pharmaceutical compounds, cosmetics, and as a bio-filler in composites (Trache et al. 2016).

PPF contains 5–6% (dry basis) of residual oil after CPO extraction. The residual oil is rich in natural carotene (4000–6000 ppm), vitamin E (2400–3500 ppm), sterols (4500–8500 ppm), coenzyme Q10 (1000–1500 ppm) (Neoh et al. 2011). Many of these bioactive compounds have found to have superior antioxidant and anticancer properties and improve human health such as reducing the risk of heart-related diseases. Attempts have been made to recover the residue oil from PPF by using solvent extraction and supercritical fluid extraction (Lau et al. 2008).

3.8 Palm Oil Mill Effluent (POME)

Besides solid wastes, the palm oil milling process also generates liquid wastes from three sources, namely sterilizer condensate, separator sludge, and hydrocyclone wastewater. These three sources have individual characteristics, when combined, forms a high-strength, acidic wastewater with high contents of organic components originating from oil, debris, fiber, cell walls, etc. An example of POME characteristics from different milling sources is as shown in Table 3.2. Out of the three sources of wastewater forming POME, separator sludge forms the majority of POME (approximately 60%) followed by sterilizer condensate (around 36%) and hydrocyclone wastewater (remaining 4%). The characteristics in Table 3.2 is just an example and could vary with improvements in milling technology and other factors that will be elaborated later in this section.

3.8.1 POME Characteristics

POME characteristics that were published from literatures are distinctly different. Other than studies conducted by Ho et al. (1984), Hwang et al. (1978), and recently Poh et al. (2010), there are barely any studies that carefully elucidate the different factors that would cause changes to POME characteristics. It is important to have a comprehensive knowledge on the range of POME characteristics to be expected as the extent of treatment would differ based on the wastewater characteristics.

In the characteristic study conducted by Poh et al. (2010), sampling of POME was conducted throughout low and high crop seasons and POME characteristics during high crop season from three different mills were studied. Based on the results in Table 3.2, pH is one parameter that remains relatively unchanged regardless of the

Table 3.2 Characteristics of POME from different milling sources (Poh et al. 2010)

Parameter (all in g/L except pH)	Sterilizer condensate	Clarification wastewater	Hydrocyclone wastewater
pH	4.5–5.5	3.5–4.5	–
Chemical oxygen demand (COD)	30–60	40–75	15
Biochemical oxygen demand (BOD)	10–25	17–35	5
Suspended solids	3–5	12–18	5–12
Total solids	40–50	35–70	5–15
Oil and grease	2–3	5–15	1–5
Total nitrogen	0.35–0.60	0.50–0.90	0.07–0.15
Ammoniacal nitrogen	0.02–0.05	0.02–0.05	–

fruiting season. Due to the higher processing capacities during high crop season, POME was found to be more concentrated and thicker compared to those discharged during a low crop season. This gives values that were evidently higher for parameters that were measured in Table 3.2. Furthermore, processing capacity and technology of the mills would also influence the characteristics of POME significantly, with higher throughput mills producing POME that is considerably more contaminated. While the palm oil industry is facing resistance from some countries due to sustainability concerns on oil palm plantation activities, it is projected that the demand of CPO in the market will remain positive. Therefore, it is crucial for mills to have adequate planning on POME treatment strategies when palm oil milling capacity will be expanded to cope with the greater demand. Planning for POME treatment in new mills should also take into consideration of movable targets; besides complying with regulations set by local environmental authorities, the mill owners should also look into the possibility of incorporating design that has a greater flexibility to accommodate changes to wastewater characteristics and also tightened regulatory standards (Table 3.3).

Since sterilizer condensate is a major source of POME, it contributes to a relatively higher discharge temperature of the wastewater. The temperature of POME measurable depends on the sampling location and retention time of POME at the oil trap prior discharge to treatment systems. Meanwhile, pH of POME is pretty consistent in the region of 4–5. The acidic characteristic is attributed to the presence of organic components and those that may have started to undergo decomposition as POME is channeled from the oil trap to treatment system.

Meanwhile, ammoniacal nitrogen (NH_3–N) and volatile fatty acid (VFA) provide an indicator on the degree of degradation of POME; where higher values of NH_3–N and VFA reflect higher degree of degradation. The degree of degradation is a valuable information for design of POME treatment system as higher degree of degradation would lead to reduced treatment time. Therefore, for cases where the location of oil trap is designed to be significantly far from the milling process and POME treatment system, it is expected that POME entering the treatment system would have undergone a significant degree of degradation, entering the treatment process with a relatively lower temperature, higher acidity, and VFA concentrations.

Referring to Table 3.2, the BOD/COD ratio of POME is greater or equals to 0.5, indicating that POME is highly biodegradable, justifying that this source of wastewater should be treated under biological means. Nevertheless, the acidity of POME can be detrimental to the treatment process. Later, the different treatment technologies for POME will be reviewed and their feasibility of implementation discussed.

3.9   POME Treatment

The conventional steps of POME treatment do not deviate significantly from any wastewater treatment process, with the block flow diagram of a conventional POME treatment system shown in Fig. 3.3. The treatment of any wastewater will generally

Table 3.3 POME characteristics sampled from three palm oil mills (Poh et al. 2010)

Parameters	Units	Low crop season (Golconda)	High crop season		
		Range	Golconda	Bukit Kerayong	Seri Ulu Langat
			Range	Range	Range
Temperature	°C	45.8–62.1	32.2–48	57.0–69.0	54–66.5
pH	–	4.4–4.5	4.32–5.1	5.18–5.50	4.18–4.7
Flow rate	m^3/h	7.8–25.4	25.7–59.6	6.69–29.61	5.4–65.0
BOD	mg/L	12520–42630	38700–85800	45000–61500	32100–56700
COD	mg/L	27840–85267	70700–75700	82100–89500	67900–87300
sCOD	mg/L	19383–41333	32300–35900	35900–37400	35600–40900
BOD/COD	–	0.45–0.76	0.53–1.15	0.54–0.69	0.41–0.68
SS	mg/L	7176–36167	18200–21450	26300–34400	24200–34300
TS	mg/L	16333–60000	39800–47000	49400–60900	51820–60940
DS	mg/L	9157–23833	20350–26000	23000–29800	25500–30440
TVS	mg/L	12000–47667	31140–36690	44740–56660	41180–47060
TP	mg/L	N/A	675–1220	1360–1770	690–910
TOC	mg/L	N/A	16700–20100	24300–44700	20200–22200
O&G	mg/L	2500–16100	4982–7914	23159–29728	11004–15880
NH_3–N	mg/L	103–210	90–248	16–26	25–67
TKN	mg/L	230–780	375–1350	563–1575	525–1350
TN	mg/L	N/A	750–1020	720–1040	920–1020
VFA (acetic acid)	mg/L	1215–2546	2742–4085	547–2052	307–861
SO_4	mg/L	N/A	0–20	0–20	0–10
Lignin	mg/L	N/A	1425–1825	1450–1525	1625–1900
Toxicity	% inhibition	N/A	(−1.5) to (−29.0)	(−33.7) to (−59.8)	(−19.1) to (−122.0)

start with a pretreatment system, followed by a primary, secondary, and a tertiary treatment step prior to discharge. However, options or combinations of technologies that can be employed for treatment of wastewater can be vastly different depending on the desired outcome and various factors (i.e., cost, speed, and effectiveness of treatment, availability of human resource).

Conventionally, POME treatment process starts from the oil trap, where the oil trap acts as a pretreatment step, to remove excessive amount of oil and grease (O&G) and in the process, some of the denser or bulkier particles are removed in this process. Subsequently, the POME from oil trap enters a series of ponds (anaerobic → facultative → aerobic), which is normally regarded as a ponding system to remove the bulk of organic materials and solids present in the wastewater to meet the regulatory

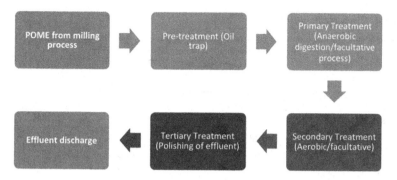

Fig. 3.3 Basic block flow diagram of POME treatment system

standards. Normally, POME will be discharged to the watercourse once the treated effluent meets the regulatory standards after primary and secondary treatment, and therefore the tertiary treatment is not present. However, tertiary treatment is added into Fig. 3.2 as there is a need to introduce tertiary treatment, which will be discussed in the subsequent section.

In the ponding system, activated sludge which consists of a mixed population of bacteria is present to degrade complex organic molecules to give water and other by-products such as biogas (methane, carbon dioxide, and hydrogen sulfide). VFA is also generated as an intermediate product from the degradation of organic molecules which becomes the food to be consumed by methane-producing bacteria. At the very beginning when the number of palm oil mills is still small, the emission of biogas and odor from VFA generation is acceptable due to the remote location of palm oil mills and limited volume of biogas generated. As the palm oil industry prospers, the number of mills increased, and some of these mills are situated near to residential area, causing the bad odor from hydrogen sulfide and VFA to affect the health and well-being of nearby residents. Furthermore, methane present in the biogas contributes to greenhouse gas emission to the environment. Hence, government authorities have put in place strict regulations and guidelines for palm oil mills to adhere to eliminate environmental and health related issues. The discharge regulations for liquid wastes from palm oil industry in Malaysia are as listed in Table 3.4. There were discussions mulling on reducing the BOD concentration to 20 mg/L for final discharge in 2015. However, amendments have not been finalized on the regulations as there are concerns on whether the mills are able to comply with the stricter guidelines.

There is a huge room to explore in terms of managing POME generated from palm oil milling. The main drivers that drive changes or improvements in POME treatment systems are (i) regulatory requirements, (ii) resource conservation (possibility of water reuse), and (iii) possible revenue from using biogas generated from POME treatment to generate electricity and supply them to the national grid. On the other hand, initial investment cost to make modifications to existing POME treatment is one of the main causes of resistance to change. Furthermore, there are a lot of uncertainties that the mills have to deal with due to the implementation of change,

Table 3.4 Discharge limits for POME

Parameter (all in g/L except pH)	Environment quality act (Prescribed Premises) (Crude Palm Oil) regulations 1977	Proposed new regulations (2015)
BOD_3	100	20
COD	–	–
Total solids	–	–
Suspended solids	400	200
Oil & grease	50	5
Ammoniacal nitrogen	150[a]	150[a]
Total nitrogen	200[a]	200[a]
pH	5–9	5–9
Temperature (°C)	45	45

[a]Value of filtered sample

where some of these uncertainties can be huge risks to small and independent mills. These will be discussed in subsequent sections in this chapter.

3.9.1 Alternative Treatment Methods

Other than the ponding system, various studies have been conducted in search of the best technology that is able to treat POME, generating consistent effluent quality that meets regulatory standards. In this section, we will be reviewing all the different treatment methods that have been employed or can be potentially employed for POME treatment. Figure 3.4 is a compilation of various treatment methods studied for POME treatment. Some of these methods have been commercialized while some remain under development in the laboratory.

(i) Biological Treatment

Biological treatment systems are most prevalent in POME treatment for the reason that POME has high concentrations of organic compounds that causes a lot of problems for a lot of other treatment systems. This technology has been used to deal with many industrial wastewaters that have high organic content and have been proven to be successful. There are two modes of biological treatment systems, namely (i) anaerobic and (ii) aerobic digestion.

Anaerobic digestion is a process where complex organic matters are broken down with the absence of oxygen to produce water, methane, and carbon dioxide, while the latter also reduces contaminants with the aid of oxygen, but does not produce biogas as one of the by-products. These two technologies are considered biological treatment processes for the reason that microorganisms are involved in the process.

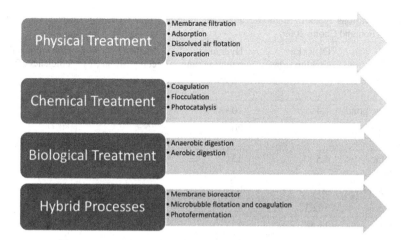

Fig. 3.4 Treatment methods studied for POME treatment

These two biological treatment processes can happen in any containment with acclimatized microorganisms and in general, are conducted in a sequence where anaerobic digestion occurs prior to aerobic digestion. At times when industrial wastewaters are less contaminated, aerobic digestion is employed individually as the rate of aerobic digestion is more rapid compared to anaerobic digestion. Ponding system employed for POME treatment is considered as an integration of both anaerobic and aerobic digestion, where some ponds are strictly anaerobic, followed by facultative ponds (with low content of oxygen) and ending with aerobic ponds. This method is preferred due to ease in operation, low capital investment, and operating cost.

However, the ponding system has low treatment efficiency, occupying large areas, causing the entire POME treatment process to last between 30 and 90 days. With increased CPO processing capacity, availability of land for expansion of ponding system becomes a pressing issue, especially if the mill is situated in hot areas with huge development potential. Therefore, it is essential for mills to look into alternatives that are able to cope with the expansion of milling activity with smaller space requirements.

Closed digesters, high-rate anaerobic digesters, and aeration tanks then became essential to address the above-mentioned issue. Continuous Stirred Tank Reactor (CSTR) was used previously for POME treatment (Tong and Jaafar 2006) but did not show obvious improvement in terms of effluent treatment quality as compared to anaerobic ponds, only achieving 80% of COD removal efficiency whereas 97.8% was achieved with anaerobic ponds. However, the treatment time and methane composition using CSTR were significantly improved, where CSTR required 18 days to achieve 80% COD removal, producing 62.5% of methane while the latter required 40 days to achieve 97.8% COD removal but only producing an average methane

Table 3.5 Various types of high-rate anaerobic digesters that were implemented for POME treatment (Poh and Chong 2009)

Types of high-rate anaerobic digester	OLR (kg COD/m^3 day)	Hydraulic retention time (days)	Methane composition (%)	COD removal efficiency (%)
Anaerobic pond	1.4	40	54.4	97.8
Anaerobic digester	2.16	20	36	80.7
Anaerobic filtration	4.5	15	63	94
Fluidized bed	40.0	0.25	N/A	78
UASB	10.63	4	54.2	98.4
UASFF	11.58	3	71.9	97
CSTR	3.33	18	62.5	80
Anaerobic contact process[a]	3.44	4.7	63	93.3

[a]In terms of BOD
N/A Data unavailable

concentration of 54.4%. Table 3.5 is a list of the other high-rate anaerobic digesters that have been researched to be implemented for POME treatment.

Data from Table 3.5 indicate that high-rate anaerobic digesters in general are able to handle higher throughput of POME compared to anaerobic pond. However, there is a trade-off with initial capital investment to implement changes. Conventional process though takes longer time to complete treatment, the process is more stable toward changes in POME characteristics. These high-rate anaerobic digesters were reported to have some issues dealing with high oil and grease and solid loading in POME.

More recently, Poh and Chong (2014) have developed an improved design of high-rate anaerobic digester (Upflow Anaerobic Sludge Blanket-Hollow Centered Packed Bed (UASB-HCPB) reactor) that is able to resolve most of the issues faced by the existing designs when dealing with POME treatment. The UASB-HCPB reactor was operated at 55 °C to harness the heat from POME to increase the rate of anaerobic digestion. This system is able to operate with a hydraulic retention time of 2 days, OLR of 27.65 g/L day, and MLVSS concentration of 14.7 g/L, removing 90% of COD and BOD, 80% of suspended solid, and at the same time, produce 60% of methane.

Technological advancement of high-rate anaerobic digesters has enabled improved treatment efficiency and also sustainability of this technology as a source of renewable energy. Nevertheless, anaerobic digestion itself is a sensitive process that can be influenced by various factors. Without skilled professionals trained to manage high-rate anaerobic digesters, palm oil mills may still face significant downtime if the high-rate anaerobic digesters become unstable due to variation in POME characteristics and the other factors that may cause instability.

Meanwhile, sequencing batch reactor, activated sludge processes, and other variations of reactor were commonly used as a follow up to polish the effluent coming out from the anaerobic digesters (Vijayaraghavan et al. 2007; Ahmed and Idris 2006; Chan et al. 2010). While the retention time of these systems has significantly reduced as compared to ponding systems, the effluent quality would have met discharge standards, the visual of the treated effluent were often not reported. In addition, there are concerns about generation of excess sludge from the aerobic process which is also an existing problem with the utilization of ponds. As such, many mills opted to retain the conventional operation due to low operating costs, without the need to inject additional capital investment.

(ii) Physical treatment

Of those that were mentioned for physical treatment in Fig. 3.4, the two more well-known technologies tested on POME treatment are evaporation and membrane filtration. The main characteristics of physical treatment are its ability to treat POME at a much shorter duration as compared to biological treatment processes.

In 2002, a Malaysian wastewater treatment company in collaboration with MPOB has implemented evaporation as a method to treat palm oil mill effluent, producing quality fertilizer. A falling film evaporator was utilized to concentrate POME by 30–40%. The condensed water was found to meet the discharge limit of POME to the watercourse, but it was reported that the appearance of the water may sometimes be slightly turbid (Wee et al. 2002).

The advantage of such system is that the concentrated solids can then be made into fertilizer. However, energy intensiveness due to the need to generate steam for the falling film evaporator as well as energy required to dry the wet solids is a disadvantage as energy conservation has to be taken into consideration (Ma 1997). Though the energy intensiveness can be brought down by incorporating more evaporators and introducing efficient thermal vapor recompression, it implies that mills have to put in a greater amount of capital investment to reduce the energy intensiveness.

Meanwhile, with many successful applications of membrane filtration for treatment of various wastewaters, similar approach has been tested to treat POME. Ahmad et al. (2003) have conducted a pilot study of a complete POME treatment system which comprised of coagulation, sedimentation, and adsorption for pretreatment, followed by ultrafiltration and reverse osmosis process in the same order to produce effluent with quality of boiler feed water. With an efficiency of 98.8% reduction of COD and 99.4% reduction of BOD, the membrane treatment process could be a good option moving forward.

On the other hand, the pretreatment process has to be highly efficient to reduce as much contaminants as possible from POME to avoid clogging of membrane. The same work from Ahmad et al. (2003) has pointed out that the membrane flux declined by 5% for ultrafiltration and 1.5% for reverse osmosis after cleaning. This implies that consistent maintenance of the membrane is essential to ensure good permeate quality from the process. Comparing to conventional ponding system, the implementation of membrane technology requires high capital investment as well as operating costs to maintain operating pressures of ultrafiltration and reverse osmosis unit which operates at a maximum pressure of 4.5 bar and 50 bar, respectively. However, should membrane with increased resilience to POME characteristics be developed, membrane technology could potentially be an effective way to treat POME for non-potable uses.

(iii) Chemical treatment

Chemical treatment of POME involved the use of chemical aids to remove contaminants such as the addition of coagulants/flocculants for pre- or primary treatment as well as the use of photocatalysis for effluent polishing. There has been an emphasis on the utilization of natural coagulants (i.e., *Moringa olifera*, unmodified rice starch, *Cassia obtusifolia*, etc.) for pretreatment of POME (Bhatia et al. 2007; Teh et al. 2014; Shak and Wu 2014) and conducting electrocoagulation for POME treatment (Agustin et al. 2008).

The use of *Moringa olifera* as a coagulant effectively removed 95% suspended solids and 52.2% of COD, having worked better at lower temperature range, indicating that POME, typically released at temperatures in the range of 80–90 °C has to be cooled prior to treatment. Meanwhile, studies by Teh et al. (2014) and Shak and Wu (2014) both indicated that natural coagulants were more effective than inorganic coagulants and that conventional coagulants such as alum can be replaced for effective treatment of POME. However, the use of these coagulants was not taken up to full speed possibly due to the cost of extraction which will not be cost-effective as compared to industrial-use coagulants.

Meanwhile, in the study on electrocoagulation of POME conducted by Agustin et al. (2008), it can be concluded that electrocoagulation would be effective on the removal of iron content from the solution, but the performance on COD and BOD removal was not impressive. Furthermore, in order to facilitate electrocoagulation, an additional step of residual oil removal from POME has to be conducted, complicating the entire treatment process.

As POME has been difficult to be effectively treated to meet discharge limits by solely relying on biological treatment processes, researchers have also ventured into the use of photocatalyst as a medium to be added into POME as a polishing step to further bring down the level of contamination. Titanium dioxide and zinc oxide were generally used as photocatalytic substances, where these materials were doped with other metals (i.e., silver, platinum, etc.) to enhance its performance (Cheng et al. 2015; Ng and Cheng 2015; Ng et al. 2016).

While photocatalyst was found to be a superior material for water splitting, producing hydrogen that can be extracted as a form of fuel from POME, the greatest challenge remains to be the implementation of an effective pretreatment process prior to polishing of POME using photocatalysts. The dark color due to melanoidins and debris contained in POME causes light not to be able to penetrate to reach the photocatalysts for the effective degradation of organic components to produce hydrogen. Furthermore, POME is a complex wastewater which contains an array of organic compounds that is hard to degrade and might hinder the process of water splitting (Ng et al. 2019). Therefore, Ng et al. (2019) have recommended to elucidate the mechanisms of photodegradation of POME in order to effectively tackle the problems attributed to POME characteristics.

(iv) Hybrid treatment

The sections which preceded were more common and established methods to conduct POME treatment. Researchers learned about the pros and cons of different technologies and developed hybrid systems, for example, membrane bioreactor, photofermentation, microbubble treatment coupled with coagulation.

While biological treatment systems are capable of removing the bulk of contaminants in the system, they do not produce effluent that is clear and meeting discharge standards as compared to membranes. On the other hand, membrane systems would require high-strength wastewater such as POME to be pretreated to avoid fouling and clogging issues. Therefore, bioreactor incorporating membrane technology was studied in different schemes. For example, Abdurahman et al. (2011) installed a membrane unit to separate the sludge suspension after anaerobic digestion, and the retentate was recycled back into the process; Ahmad et al. (2011) on the other hand, treated POME through a series of anaerobic, anoxic, and aerobic reactor, where the membrane unit was immersed into the aerobic reactor to produce clean permeate.

As the study conducted by Ahmad et al. (2011) involved more steps prior to feeding POME into the membrane unit, the permeate was much cleaner, achieving an average COD of 106 mg/L and suspended solid concentration of 16 mg/L. However, it is evident that the performance deteriorated with membrane fouling. Hence, it is imperative to focus on the elimination of fouling issues when using membrane for POME treatment.

Biological treatment of POME has mostly been conducted with mixed bacteria culture for the production of biogas, with methane as the main component. Meanwhile, hydrogen, a much cleaner fuel could also be produced via fermentation. However, if a mixed culture is used, the fermentation condition has to be carefully controlled to suppress the production of other unwanted by-products. Therefore, specific species of bacteria that requires light source for activation, specifically purple non-sulfur bacteria, were studied and implemented for POME treatment to produce biohydrogen.

Budiman et al. (2015) have studied the use of *Rhodobacter sphaeroids* NCIMB8253 to treat POME for biohydrogen production. However, the dark color and turbidity of POME do not allow effective conversion of biohydrogen and this problem is alleviated with the addition of pulp and paper mill effluent as a diluting agent to increase light penetration during the process. The maximum biohydrogen production can be obtained only with 25% v/v of POME at 4.67 mL H_2/ml medium. A separate study by Jamil et al. (2009) indicated that photofermentation of POME also has a limitation, causing relatively lower biohydrogen production. This is because purple non-sulfur bacteria require organic acids to produce biohydrogen and this is not abundantly available in raw POME. As such, pretreatment processes would be

Table 3.6 Performance of microbubble treatment on anaerobically digested POME (Poh et al. 2014)

Parameters	Anaerobically digested POME (ADPOME)	Effluent after microbubble flotation (19.8 L/min)	Effluent after microbubble flotation (19.8 L/min) + PAC coagulation
pH[a]	7.05	8.06	6.28
Temperature (°C)	18	26	25
COD (mg/L)	21025	9725	1407
BOD (mg/L)	2220	510	Not detected
TSS (mg/L)	17995	7685	22
O&G (mg/L)	235	60	Not detected

[a]pH has no units

important to breakdown the complex molecules to organic acids to facilitate effective biohydrogen production.

Microbubble flotation, where air bubbles were produced in the micron size range has also been implemented in the replacement of aerobic digestion for POME treatment (Poh et al. 2014). This is because flotation could remove odor, decolorising of wastewater as well as recovery of useful materials which many other physical/chemical processes could not achieve simultaneously (Cappono et al. 2006). Furthermore, the use of microbubble flotation as an alternative to aeration could possibly eliminate the production of sludge.

In the study conducted by Poh et al. (2014), microbubbles were generated via ejector type venturi device. Table 3.6 lists the results from the microbubble study. The system was capable of removing 53.7% of COD, 77.0% of BOD, and managed a removal of 74.5% of oil and grease just with 10 min of retention time. However, the removal of total suspended solids was not satisfactory. Therefore, coagulant (polyalumimun chloride (PAC)) was induced into the microbubble flotation system, reducing 99.7% of the suspended solid, producing clear solution. BOD and oil and grease were undetected after the microbubble treatment with coagulant, indicating that organic matter in anaerobically digested POME was effectively removed. The promising results from this study indicated that there is a possibility of producing clean effluent by replacing aerobic digestion with microbubble treatment process. This is however possible provided that the anaerobic digester produces consistent effluent to the microbubble system.

3.10 Conclusion

In this chapter, we reviewed the palm oil milling process. Solid wastes are generated via the stripping, nut/fiber separation, nut drying, and winnowing, producing wastes such as EFB, PKS, and PPF. These wastes were typically incinerated where ashes from EFB can be used as fertilizer while the incineration of PKS and PPF would produce heat and energy to sustain mill operation. Other methods derived to deal with solid wastes are such as the conversion of wastes into bio-oil or biochar via pyrolysis, extraction of cellulosic material from fibers, and some of these wastes were made into briquettes to support energy generation in other industries. While there is an advantage of extracting valuable products from solid wastes, some of the processes are complicated and require huge capital investment, deterring implementation.

On the other hand, palm oil mills also produce wastewater (POME) from sterilization, hydrocyclone and centrifugation process in the production of CPO. The wastewater generated is characterized to be acidic and highly polluting with a huge amount of organic compounds and solids entrained from the milling process. Generally treated with a series of ponds, there are still issues to be dealt with the treatment of POME; e.g., the release of greenhouse gasses to the environment, occupying large land areas, and low treatment efficiency with long retention time. Moreover, the inconsistency of POME characteristics which changes according to load, season as well as fruit types provides a challenge to develop a system that is fully robust to deal with the variability of this wastewater. While there are many technologies evaluated and found suitable, these were evaluated on an individual level and not considering the entire chain of treatment processes. There is a need to conduct further investigation to evaluate the robustness of the combination of different technologies that are applicable for POME treatment to effectively treat the wastewater, meeting discharge standards, conserving not just the environment but also water that is precious to sustain lives.

References

Abduraman, N. H., Rosli, Y. M., & Azhari, N. H. (2011). Development of a membrane anaerobic system (MAS) for palm oil mill effluent (POME) treatment. *Desalination, 266*, 208–212.

Abnisa, F., et al. (2013). Characterization of bio-oil and bio-char from pyrolysis of palm oil wastes. *BioEnergy Research, 6*(2), 830–840.

Abu Bakar, R., et al. (2011). Effects of ten year application of empty fruit bunches in an oil palm plantation on soil chemical properties. *Nutrient Cycling in Agroecosystems, 89*(3), 341–349.

Agustin, M. B., Sengpracha, W. P., & Phutdhawong, W. (2008). Electrocoagulation of palm oil mill effluent. *International Journal of Environmental Research and Public Health, 5*, 177–180.

Ahmad, A., Buang, A., & Bhat, A. H. (2016). Renewable and sustainable bioenergy production from microalgal co-cultivation with palm oil mill effluent (POME): A review. *Renewable and Sustainable Energy Reviews, 65*, 214–234.

Ahmed, M., & Idris, A. (2006). Effects of organic loading on performance of aerobic fluidized using diluted palm oil mill effluent. *Suranaree Journal of Science and Technology, 13*, 299–306.

Ahmad, A. L., Ismail, S., & Bhatia, S. (2003). Water recycling from palm oil mill effluent (POME) using membrane technology. *Desalination, 157*, 87–95.

Ahmad, Z., Ujang, Z., Olsson, G., & Abdul Latif, A. A. (2011). Development of a membrane anaerobic system (MAS) for palm oil mill effluent (POME) treatment. *International Journal of Integrated Engineering, 1*(2), 17–25.

Awalludin, M. F., et al. (2015). An overview of the oil palm industry in Malaysia and its waste utilization through thermochemical conversion, specifically via liquefaction. *Renewable and Sustainable Energy Reviews, 50,* 1469–1484.

Bhathia, S., Othman, Z., & Ahmad, A. L. (2007). Pretreatment of palm oil mill effluent (POME) using moringa olifera seeds as natural coagulant. *Journal of Hazardous Materials, 145,* 120–126.

Budiman, P. M., Wu, T. Y., Ramanan, R. N., & Md Jahim, J. (2015). Improvement of biohydrogen production through combined reuses of palm oil mill effluent together with pulp and paper mill effluent in photofermentation. *Energy & Fuels, 29,* 5816–5824.

Cappono, F., Sartori, M., Souza, M. L., & Rubio, J. (2006). Modified column flotation of adsorbing iron hydroxide colloidal precipitates. *International Journal of Mineral Processing, 79,* 167–173.

Chan, Y. H., et al. (2014). Bio-oil production from oil palm biomass via subcritical and supercritical hydrothermal liquefaction. *The Journal of Supercritical Fluids, 95,* 407–412.

Chan, Y. J., Chong, M. F., & Law, C. L. (2010). Biological treatment of anaerobically digested palm oil mill effluent (POME) using a Lab-Scale Sequencing Batch Reactor (SBR). *Journal of Environmental Management, 91,* 1738–1746.

Cheng, C. K., Derahman, M. R., & Khan, M. R. (2015). Evaluation of the photocatalytic degradation of pre-treated palm oil mill effluent (POME) over Pt-loaded titania. *Journal of Environmental Chemical Engineering, 3,* 261–270.

Dhyani, V., & Bhaskar, T. (2018). A comprehensive review on the pyrolysis of lignocellulosic biomass. *Renewable Energy, 129,* 695–716.

Fahma, F., et al. (2010). Isolation, preparation, and characterization of nanofibers from oil palm empty-fruit-bunch (OPEFB). *Cellulose, 17*(5), 977–985.

Hassan, T.M., et al. (2019). Optimizing the acid hydrolysis process for the isolation of microcrystalline cellulose from oil palm empty fruit bunches using response surface methods. *Waste and Biomass Valorization.*

Hastuti, N., Kanomata, K., & Kitaoka, T. (2018). Hydrochloric acid hydrolysis of pulps from oil palm empty fruit bunches to produce cellulose nanocrystals. *Journal of Polymers and the Environment, 26*(9), 3698–3709.

Ho, C. C., Tan, Y. K., & Wang, C. W. (1984). The distribution of chemical constituents between the soluble and the particulate fractions of palm oil mill effluent and its significance on its utilization/treatment. *Agricultural Wastes, 11,* 61–71.

Husain, Z., Zainac, Z., & Abdullah, Z. (2002). Briquetting of palm fibre and shell from the processing of palm nuts to palm oil. *Biomass and Bioenergy, 22*(6), 505–509.

Hwang, T. K., Ong, S. M., Seow, C. C., & Tan, H. K. (1978). Chemical composition of palm oil mill effluents. *Planter, 54,* 749–755.

Irvan. (2018). Processing of palm oil mill wastes based on zero waste technology. *IOP Conference Series: Materials Science and Engineering, 309,* 012136.

Jamil, Z., Annuar, M. S. M., Ibrahim, S., & Vikineswary, S. (2009). Optimisation of phototrophic hydrogen production by Rhodopseudomonas palustris PBUM 001 via statistical experimental design. *International Journal of Hydrogen Energy, 34,* 7502–7512.

Kong, S.-H., et al. (2014). Biochar from oil palm biomass: A review of its potential and challenges. *Renewable and Sustainable Energy Reviews, 39,* 729–739.

Lau, H. L. N., et al. (2006). Quality of residual oil from palm-pressed mesocarp fiber (Elaeis guineensis) using supercritical CO2 with and without ethanol. *Journal of the American Oil Chemists' Society, 83*(10), 893–898.

Lau, H. L. N., et al. (2008). Selective extraction of palm carotene and vitamin E from fresh palm-pressed mesocarp fiber (Elaeis guineensis) using supercritical CO2. *Journal of Food Engineering, 84*(2), 289–296.

Lehmann, J., Gaunt, J., & Rondon, M. (2006). Bio-char sequestration in terrestrial ecosystems—A review. *Mitigation and Adaptation Strategies for Global Change, 11*(2), 403–427.

Liew, W. L., Kassim, M. A., Muda, K., Loh, S. K., & Affam, A. C. (2015). Conventional methods and emerging wastewater polishing technologies for palm oil mill effluent treatment: A review. *Journal of Environmental Management, 149,* 222–235.

Lim, K. C., & Rahman, Z. A. (2002). The effects of oil palm empty fruit bunches on oil palm nutrition and yield, and soil chemical properties. *Journal of Oil Palm Research, 14*(2), 1–9.

Ma, A. N. (1997). Evaporation technology for pollution abatement in palm oil mills. *1997 National Seminar on Palm Oil Milling, Refining Technology and Quality* (pp. 167–170).

Mahlia, T. M. I., et al. (2001). An alternative energy source from palm wastes industry for Malaysia and Indonesia. *Energy Conversion and Management, 42*(18), 2109–2118.

Mohammad, N., et al. (2012). Effective composting of oil palm industrial waste by filamentous fungi: A review. *Resources, Conservation and Recycling, 58*, 69–78.

Moradi, A., et al. (2015). Effect of four soil and water conservation practices on soil physical processes in a non-terraced oil palm plantation. *Soil and Tillage Research, 145*, 62–71.

Nasri, N. S., et al. (2014). Assessment of porous carbons derived from sustainable palm solid waste for carbon dioxide capture. *Journal of Cleaner Production, 71*, 148–157.

Nasrin, A. B., et al. (2008). Oil palm biomass as potential substitution raw materials for commercial biomass briquettes production. *American Journal of Applied Sciences, 5*(3), 179–183.

Neoh, B., et al. (2011). Palm pressed fibre oil: A new opportunity for premium hardstock? *International Food Research Journal, 18*, 769–773.

Ng, K. H., & Cheng, C. K. (2015). A novel photomineralization of POME over UV-responsive TiO2 photocatalyst: Kinetics of POME degradation and gaseous product formations. *RSC Advances, 65*, 53100–53110.

Ng, K. H., Lee, C. H., Khan, M. R., & Cheng, C. K. (2016). Photocatalytic degradation of recalcitrant POME waste by using silver doped titania: Photokinetics and scavenging studies. *Chemical Engineering Journal, 286*, 282–290.

Ng, K. H., Lai, S. Y., Cheng, C. K., Chen, K., & Fang, C. (2019). TiO2 and ZnO photocatalytic treatment of palm oil mill effluent (POME) and feasibility of renewable energy generation: A short review. *Journal of Cleaner Production, 233*, 209–225.

Onoja, E., et al. (2018). Oil palm (Elaeis guineensis) Biomass in Malaysia: The present and future prospects. *Waste and Biomass Valorization*.

Poh, P. E., & Chong, M. F. (2009). Development of anaerobic digestion methods for palm oil mill effluent (POME) treatment. *Bioresource Technology, 100*, 1–9.

Poh, P. E., Yong, W. -J., & Chong, M. F. (2010). Palm oil mill effluent (POME) characteristics in high crop season and the applicability of high-rate anaerobic bioreactors for the treatment of POME. *Industrial & Engineering Chemistry Research, 49*, 11732–11740.

Poh, P. E., & Chong, M. F. (2014). Upflow anaerobic sludge blanket-hollow centered packed bed (UASB-HCPB) reactor for thermophilic palm oil mill effluent (POME) treatment. *Biomass and Bioenergy, 67*, 231–242.

Poh, P. E., Ong, J. W. Y., Lau E. V., Chong, M. N. (2014). Investigation on micro-bubble flotation and coagulation for the treatment of anaerobically treated palm oil mill effluent (pome). *Journal of Environmental Chemical Engineering, 2*, 1174–1181.

Prasertsan, S., & Prasertsan, P. (1996). Biomass residues from palm oil mills in Thailand: An overview on quantity and potential usage. *Biomass and Bioenergy, 11*(5), 387–395.

Rangabhashiyam, S., & Balasubramanian, P. (2019). The potential of lignocellulosic biomass precursors for biochar production: Performance, mechanism and wastewater application—A review. *Industrial Crops and Products, 128*, 405–423.

Sabil, K. M., et al. (2013). Effects of torrefaction on the physiochemical properties of oil palm empty fruit bunches, mesocarp fiber and kernel shell. *Biomass and Bioenergy, 56*, 351–360.

Salètes, S., et al. (2004). Ligno-cellulose composting: Case study on monitoring oil palm residuals. *Compost Science & Utilization, 12*(4), 372–382.

Samiran, N. A., et al. (2016). Progress in biomass gasification technique—With focus on malaysian palm biomass for syngas production. *Renewable and Sustainable Energy Reviews, 62*, 1047–1062.

Shak, K. P. Y., & Wu, T. Y. (2014). Coagulation-flocculation treatment of high-strength agro-industrial wastewater using natural cassia obtusifolia seed gum: Treatment efficiencies and floc characterization. *Chemical Engineering Journal, 256*, 293–305.

Sharma, A. K., et al. (2015). Pilot scale soda-anthraquinone pulping of palm oil empty fruit bunches and elemental chlorine free bleaching of resulting pulp. *Journal of Cleaner Production, 106*, 422–429.

Siddiquee, S., Shafawati, S. N., & Naher, L. (2017). Effective composting of empty fruit bunches using potential trichoderma strains. *Biotechnology Reports, 13*, 1–7.

Singh, P., et al. (2013). Using biomass residues from oil palm industry as a raw material for pulp and paper industry: Potential benefits and threat to the environment. *Environment, Development and Sustainability, 15*(2), 367–383.

Singh, R. P., et al. (2010). Composting of waste from palm oil mill: A sustainable waste management practice. *Reviews in Environmental Science and Bio/Technology, 9*(4), 331–344.

Sukiran, M. A., et al. (2017). A review of torrefaction of oil palm solid wastes for biofuel production. *Energy Conversion and Management, 149*, 101–120.

Sulaiman, F., et al. (2011). An outlook of Malaysian energy, oil palm industry and its utilization of wastes as useful resources. *Biomass and Bioenergy, 35*(9), 3775–3786.

Teh, C. Y., Wu, T. Y., & Juan, J. C. (2014). Optimisation of agro-industrial wastewater treatment using unmodified rice starch as a natural coagulant. *Industrial Crops and Products, 56*, 17–26.

Thambirajah, J.J., Zulkali, M.D., & Hashim, M.A. (1995). Microbiological and biochemical changes during the composting of oil palm empty-fruit-bunches. Effect of nitrogen supplementation on the substrate. *Bioresource Technology, 52*(2), 133–144.

Then, Y. Y., et al. (2013). Oil palm mesocarp fiber as new lignocellulosic material for fabrication of polymer/fiber biocomposites. *International Journal of Polymer Science, 2013*, 7.

Tong, S. L., & Jaafar, A. B. (2006). POME Biogas capture, upgrading and utilization. *Palm Oil Engineering Bulletin, 78*, 11–17.

Trache, D., et al. (2016). Microcrystalline cellulose: Isolation, characterization and bio-composites application—A review. *International Journal of Biological Macromolecules, 93*, 789–804.

Vijayaraghavan, K., Ahmad, D., & Abdul Aziz, M. E. (2007). Aerobic treatment of palm oil mill effluent. *Journal of Environmental Management, 82*, 24–31.

Wan Daud, W.R., & Law, K. N. (2010). Oil palm fibers as papermaking material: Potentials and challenges. *BioResources, 6*.

Wee, V., Tan, T., Keng, P., Chua, S.T., & Ma, A. N. (2002). Closed loop zero waste system for palm oil mill. *2002 National Seminar on Palm Oil Milling, Refining Technology, Quality and Environment*. Kota Kinabalu, Sabah.

Wu, T. Y., Mohammad, A. W., Jahim, J. M., & Anuar, N. (2010). Pollution control technologies for the treatment of palm oil mill effluent (POME) through end-of-pipe processes. *Journal of Environmental Management, 91*(7), 1467–1490.

Yahya, A., et al. (2010). Effect of adding palm oil mill decanter cake slurry with regular turning operation on the composting process and quality of compost from oil palm empty fruit bunches. *Bioresource Technology, 101*(22), 8736–8741.

Chapter 4
High-Rate Anaerobic Digestion of POME for Stable Effluent and Biogas Production

In the previous chapter, we identified various technologies that can be implemented for POME treatment. Meanwhile, this chapter covers a case study focused specifically on the use of high-rate anaerobic digester for POME treatment and improvements to the process with the aim to produce stable effluent and biogas production from anaerobic digestion.

Anaerobic digestion is the focus and can be considered as the heart of POME treatment for the following reasons:

- POME is a high-strength wastewater with a lot of organic matters. It has great value to be converted to a renewable energy source while producing intermediate effluent that can be better managed.
- Anaerobic digestion is more versatile to handle a larger range of POME load compared to other technologies such as membrane filtration and photocatalysis degradation.
- Anaerobic digestion is also less energy-intensive and produces less sludge compared to aerobic digestion which requires energy for aeration.

4.1 Upflow Anaerobic Sludge Blanket-Hollow Centered Packed Bed (UASB-HCPB) Reactor

The high-rate anaerobic digester that will be elaborated in this chapter is a UASB-HCPB reactor (Poh and Chong 2014). The UASB-HCPB reactor (Fig. 4.1) is a hybrid design developed to combine the positive attributes of an upflow anaerobic sludge blanket (UASB) reactor and those of a packed bed reactor as follows:

- UASB reactor is able to facilitate the formation of granules with good settling property which in turn will provide higher retention of biomass in the system. Having higher solid retention will be desirable to increase the contact of microbes and substrate, hence improving the production of biogas and effluent quality.

© Springer Nature Switzerland AG 2020
P. E. Poh et al., *Waste Management in the Palm Oil Industry*,
Green Energy and Technology, https://doi.org/10.1007/978-3-030-39550-6_4

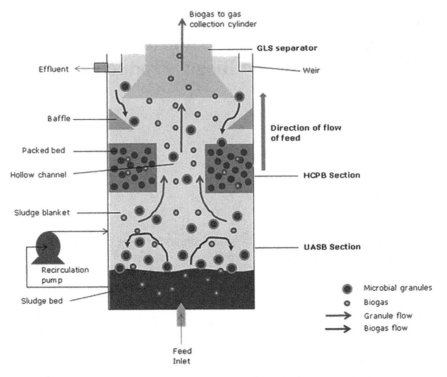

Fig. 4.1 UASB-HCPB reactor configuration (Poh and Chong 2014)

- Packed bed reactor on the other hand is supposed to aid in the immobilization of suspended biomass to reduce sludge washout and the volume required for sludge settling after anaerobic digestion. The immobilization of suspended biomass also allows greater biomass surface for anaerobic digestion.

Despite the advantages of UASB and packed bed reactor to retain biomass for anaerobic digestion, packed bed reactors can clog easily at high suspended solids loading, which will be the case for POME treatment as the wastewater has huge amounts of solids. To alleviate such issues when POME with high suspended solids loading is fed into an anaerobic digester, a cylindrical channel has been introduced int the middle of the packed bed.

The introduction of such a channel enables the immobilization of microbes on the packed bed while allowing smooth flow of effluent and biogas out from the reactor which also reduces the tendency of gas bubble entrapment within the packing. Meanwhile, it is anticipated that the reduction of the packing area would potentially lead to less maintenance since the packed bed will be less susceptible to clogging.

The UASB-HCPB reactor was started up with a thermophilic seed sludge that was used for POME treatment. The reactor was operated under thermophilic conditions since POME is discharged at high temperature (80–90 °C), making additional heating unnecessary while enabling the treatment of wastewater at higher reaction rates. This

Table 4.1 Startup performance of UASB-HCPB reactor compared to other anaerobic reactors in the literature for POME treatment (Poh 2012)

Reactor type[a]	Startup period (day)	HRT (day)	Operating temperature (°C)	OLR (g COD/L day)	COD removal efficiency (%)	Methane (%)
UASB-HCPB	36	1.5	55	28.12	88	52
AF	56	10.0	55	10.9	88	–
CSTR	50	25.0	55	–	–	–
UASB	40	4.0	35	1.27	90	62
UASFF	26	1.5	38	23.15	85	62
AF	105	N/A	35	–	81–94.5	60
EGSB	150	3.0	35	11.5	92	–

[a]*AF* Anaerobic Filter, *CSTR* Continuous Stirred-Tank Reactor, *UASB* Upflow Anaerobic Sludge Blanket, *UASFF* Upflow Anaerobic Sludge-Fixed Film, *EGSB* Expanded Granular Sludge Bed

will rectify the problem of existing ponding systems which requires long retention times and large treatment areas for POME treatment.

The UASB-HCPB reactor startup took 36 days, achieving an operating organic loading rate (OLR) of 28.22 g COD/L day and a consistent COD and BOD removal efficiency of 88% and 93%, respectively. When compared against other anaerobic reactors for POME treatment as shown in Table 4.1, the duration of startup for the UASB-HCPB reactor was significantly shorter but still 10 days longer than what is taken by upflow anaerobic sludge fixed film (UASFF) operated under mesophilic conditions (Poh 2012). This was attributed to the biomass washout during the process when a shock load was imposed on the UASB-HCPB reactor which has subsequently prolonged the startup as it took time to stabilize the biomass concentration in the reactor to produce consistent effluent quality and biogas production.

However, the UASB-HCPB reactor had the ability to tolerate much higher loads as compared to the other reactors during startup contributed by the HCPB section's ability to reduce the possibility of gas hold-up and reduced clogging issues of the packed bed.

After a successful startup, the UASB reactor was subjected to a series of studies to investigate the effect of several important operating parameters on the performance of the reactor and also to identify a set of parameters that are optimized to meet performance benchmark as shown in Table 4.2. The operating parameters were optimized to be able to achieve treatment efficiencies above 90% as the wastewater should be rendered as much as possible to reduce the load of the secondary treatment process or to possibly achieve the elimination of tertiary treatment to achieve treated effluent that can be discharged or reused for non-potable activities. The optimized operating conditions were found to be OLR of 27.65 g COD/L day, HRT of 2 days, and a mixed liquor volatile suspended solid (MLVSS) of 14,700 mg/L.

Table 4.2 Target response for UASB-HCPB reactor optimization (Poh 2012)

Response	Unit	Limits
COD removal efficiency	%	>90
BOD removal efficiency	%	>90
SS removal efficiency	%	>80
Methane concentration	%	>60
Methane yield	$l\ CH_4/g\ COD_{removed}$ day	>0.3

The results from optimization indicated that the UASB-HCPB reactor has the ability to operate at a much higher load than other high-rate anaerobic reactors under thermophilic conditions and a turnover rate of 2 days which enables more POME to be processed in the same duration taken to treat using conventional ponding system. Nevertheless, the reality is POME is never going to be fed into a reactor with the same properties. The characteristics of POME vary with peak season, the milling technology, and fruit types processed (Poh et al. 2010). And so, the reactor will have difficulties operating steadily even if the operating parameters were kept constant, ignoring any possible reactor overload. This will also imply that the effluent quality produced would not be consistent, and the same for biogas production.

When biogas production cannot be kept consistent, this will be an issue for mills that intend to explore electricity production or generate other uses for methane that is generated from anaerobic digestion of POME. The biogas treatment and storage units cannot be properly designed without a good indicator of the amount of biogas generated and the electricity production capacity also cannot be forecasted with a suitable level of accuracy. There is also an insecurity issue over the stability of operating a high-rate anaerobic reactor under thermophilic conditions as thermophilic reactors (operated at 42–60 °C) tend to be sensitive to changes in operating conditions. Moreover, skilled personnel to operate, monitor, and control anaerobic reactors are lacking in the mills, which causes slow feedback to address any potential instability issues—an important factor deterring the implementation of high-rate anaerobic reactors in the palm oil milling sector.

4.2 Predictive Model for Monitoring and Control of High-Rate Anaerobic Reactors

To encourage and convince industry players to implement high-rate anaerobic reactors for POME treatment, it is crucial to resolve the problem of reactor stability and find ways to overcome the lack of skilled personnel to supervise the operation of the reactor. This is crucial due to dynamic variation in POME characteristics where close monitoring and speedy feedback are essential to ensure the stability of the anaerobic reactor. In line with the effort for Malaysian industries to move toward Industrial Revolution 4.0 (IR 4.0), automation of the process would be the key to resolve this issue and the development of a prediction model for the anaerobic reactor will be the first step toward IR 4.0. Moreover, a prediction model can also be used as a tool to simulate and study changes to certain process parameters prior to implementation in the plant. Training can also be conducted with the model. This will reduce the time and costs required for pilot trials and training significantly while providing aid to decision-making prior implementation of proposed solutions in the industrial scale.

As anaerobic digestion involves huge numbers of reaction pathways, where some of them may be unknown due to the gap of information on bacteria population in the reactor as well as interaction between various species are not well known, utilization of meta-heuristics model will not be effective to model the process as meta-heuristic modeling requires significantly clear correlations and sufficient information for accurate prediction. On the other hand, artificial intelligence (AI) can be more suited to model anaerobic digestion as they are flexible and robust, with high predictive capability and ability to handle nonlinearity (Tan et al. 2018). AI-based models can also generalize the complex input–output relationship which is desirable for anaerobic digestion since there are huge numbers of reaction pathways involved due to the interaction of a mixed culture in the reactor.

Tan et al. (2018) developed a predictive model with an adaptive neural-fuzzy inference system (ANFIS) with the aim to improve the stability of the UASB-HCPB anaerobic reactor for POME treatment under thermophilic condition. ANFIS is a combination of artificial neural network (ANN) and fuzzy logic (FL), leveraging the merits while complementing the disadvantages of respective models (Atmaca et al. 2001). This ANN and FL combination has the ability to learn and handle large data sets, eliminating the requirement for a comprehensive understanding of the behavior of anaerobic digestion of POME. The time taken for training, modification as well as adaptation of the model can be significantly reduced with the development of the ANFIS model (Noori et al. 2009).

The ANFIS predictive model was constructed with four sources of input, namely pH, COD, total suspended solids (TSS), and OLR of the influent to predict the pH, COD, and TSS of the effluent from anaerobic digestion. The predictive model was developed under the pretext that it can be easily integrated into an existing anaerobic reactor with minimal modifications and the model can be easily trained using existing data that is commonly extracted for monitoring of wastewater treatment processes. Moreover, the selected variables were important for the reasons below (Tan et al. 2018):

1. pH, COD, and TSS are important parameters to describe the condition of the reactor. Meanwhile, these parameters are prescribed in government regulatory standards which have to be met for effluent discharge to a watercourse.
2. The turnover of analysis time for these parameters is short and it is relatively easy to install online sensors to continuously monitor the condition of the reactor. Most of the wastewater treatment plants would have included pH meter, for example, as part of their online monitoring. This would allow proper and timely corrective measures to be taken should the reactor experience undesirable situation (Dellana and West 2009).

Table 4.3 Errors and standard deviations for prediction of effluent pH, COD, and TSS using ANFIS

Predicted effluent output	Error (%)	Standard deviation (%)
pH	2.06	1.68
COD	8.32	7.65
TSS	6.93	6.28

3. Using fewer variables for modeling of the anaerobic reactor can reduce operating costs and initial investment due to fewer sensors used.

On top of the four inputs, historical data from the effluent were also incorporated as a feedback to the ANFIS model as these data will replace the biological and physicochemical reaction knowledge within the reactor to improve prediction. In addition, the interaction between parameters and effluent conditions would be taken into consideration by the model. Experimental data was collected across 282 days, and 214 days of data was used for training with the remaining used for model validation. The predictive model was capable of predicting the effluent pH, COD, and TSS with errors and standard deviations as presented in Table 4.3.

With satisfactory predictive ability using ANFIS, the models to predict effluent pH, COD, and TSS were incorporated into the UASB-HCPB reactor control scheme to automate the monitoring and control of the anaerobic reactor on thermophilic POME treatment at a low cost (Tan et al. 2019). The control scheme involves the manipulation of POME OLR fed into the reactor as well as regulation of pH via the dosing of alkaline. The predicted output from ANFIS will provide a command to adjust OLR by (i) reduction of OLR if COD removal efficiency declines below 80% and (ii) OLR increase if COD removal efficiency exceeds 95%. Meanwhile, alkaline will be dosed into the feed tank should the effluent pH be estimated to plunge below 6.5.

Shock loadings were imposed gradually onto the system to test the capability of the predictive model to aid with the reactor control and also the proposed control scheme. The shocks were carried out as shown in the red circles with arrows in Fig. 4.2. Though COD removal efficiency declines for shocks imposed, the results indicated that the system required only 3 days to resume normal operation and the biogas quality was relatively consistent throughout the induction of shock. However, the biogas production quantity declined with the increment in OLR. This indicates that the reactor does not have sufficient microbes to sustain the conversion of methane gas. As such, operators can closely monitor the biogas volume to decide on the need to increase sludge concentration in the anaerobic reactor.

Finally, with the aid of the control scheme as well as the predictive model, the UASB-HCPB reactor could handle loads as high as OLR of 134.78 g COD/L day. However, the quantity of the biogas would be compromised and induces stress to the microbial population in the anaerobic reactor. The highest achievable OLR without compromising reactor performance would be 39.56 g COD/L day (Table 4.4). In comparison to a UASB-HCPB reactor with manual control, the implementation of the predictive model and control scheme showed significant improvement in terms of

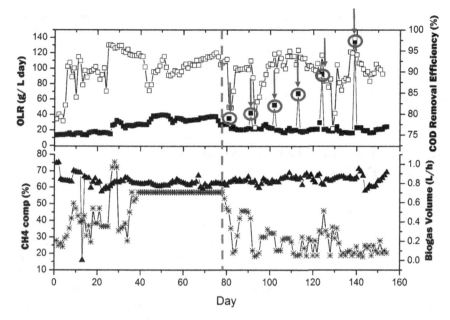

Fig. 4.2 OLR (- ■-), COD removal efficiency (-□-), methane composition (- ▲-), and biogas flowrate (-*-) profiles of the UASB-HCPB reactor for thermophilic POME treatment with implementation of the predictive model and automated control

Table 4.4 Result comparison before and after implementation of the predictive model and control scheme

Anaerobic treatment	OLR (kg/m³ day)	COD removal efficiency (%)	Methane composition (%)	HRT (Days)	References
UASB-HCPB with predictive model and control scheme	39.56	92.74	60.9	2	Tan et al. (2019)
UASB-HCPB	27.65	90.00	60.0	2	Poh and Chong (2014)

its capability to tolerate higher OLR and improved COD removal efficiency. Hence, it is imperative for mills to automate the high-rate anaerobic reactors to achieve higher performance at no expense of the need to train more skilled personnel to mend the plant. In the future, a graphical user interface will be developed to enable remote monitoring and control of anaerobic digesters since there are many mills that are still located in remote areas that are difficult to access. This will allow monitoring of anaerobic reactors in multiple locations by a single person, reducing the need for man power while retaining the effectiveness of the wastewater treatment process. The beauty of the predictive model and the control scheme described above is the adaptability of the system without major modification and investment costs, making it possible for other anaerobic reactors treating high-strength wastewater to be able to implement the same to ensure the stability of the reactor.

4.3 Pretreatment to Enhance Anaerobic Digestion of POME

With the introduction of a predictive model to a control scheme, the anaerobic digestion of POME using high-rate anaerobic reactor had an elevated performance. While the effluent production and biogas quality can now be controlled, the feed entering the high-rate anaerobic reactor is still inconsistent and may cause stress to the reactor. Besides containing water, POME is known to be constituted by complex carbohydrate, protein, and fat molecules as well as fibrous materials from the fruits. These components can be difficult to tackle and would require more time for microbes to breakdown for effective conversion to methane.

The anaerobic reactors are generally operated under conditions that favor the growth of methanogens to enable high methane conversion rate since these microbes grow at a relatively slow pace. On the other hand, the population of microbes for hydrolysis (breaking down of complex molecules) and production of volatile fatty acids will be suppressed as their growth are more favorable under acidic conditions. As anaerobic digestion works similar to a human stomach, we can imagine how consuming a lot of food that is not properly ground into suitable sizes can do to your stomach—indigestion. The same situation would actually occur to the reactors operated under high OLR and short hydraulic retention time—the microbes do not have sufficient time and condition to perform efficiently.

As such, pretreatment is thought to be essential to mediate such situations to allow the high-rate anaerobic reactors to accommodate higher loads from POME treatment. Note that in the study where shock loading was applied, the OLR of POME could reach as high as 134.78 g COD/L day and this feed was undiluted POME that came straight from the mill, indicating that there is definitely a need to preprocess the feed prior to anaerobic digestion.

Conventionally, POME produced from the mill would be directed to a cooling pond. The purpose of the cooling pond is to allow recovery of residual oil and reduction of wastewater temperature prior to pumping POME into respective ponds for further treatment. The diversion of POME into a cooling pond is justified to protect the pumps, but the heat which can be used for thermophilic anaerobic digestion is lost during the cooling process. Meanwhile, switching to a more effective pretreatment process other than a cooling pond could effectively render POME fed into the high-rate anaerobic reactor, reducing the stress of anaerobic digestion and on subsequent treatment processes. Furthermore, the ability to retain the heat from POME allows more rapid conversion of the substrate to produce biogas.

Khadaroo et al. (2019a) reviewed various pretreatment processes that could be implemented on POME treatment to eliminate the use of a cooling pond. They found that thermal pretreatment is the most promising method for POME treatment due to the high discharge temperature of POME. The POME discharge temperature at 80–90 °C requires less energy for the wastewater to reach the optimal condition for thermal pretreatment. In addition, the increase in biogas production from the anaerobic reactor as a result of feeding thermally pretreated POME could easily compensate for the energy consumption required for thermal pretreatment.

A thickening unit was proposed by Khadaroo et al. (2019b) to compensate for the capability of the cooling pond to separate residual oil and at the same time separate the solids and liquid layer of POME for thermal pretreatment. The separation of the solid and liquid layers is postulated to allow control of the organic load into the reactor. The outcome of this study indicated that hot POME is more compressible, making it more advantageous for the thickener to operate with hot POME. While this work is still ongoing, it is expected that the integration of pretreatment system with the existing automated UASB-HCPB reactor will be able to handle POME treatment efficiently and effectively to produce quality effluent for subsequent polishing processes. At the same time, the improvement of the anaerobic digestion process will allow the system to produce a sustainable amount of biogas for energy generation.

4.4 New Concept for POME Treatment Process for the Future and Possible Advancements in the Field

(i) The new concept

Drawing the success of the technological improvements described previously, the following process (Fig. 4.3) is a concept proposed to replace the conventional POME treatment process to produce treated effluent that would meet discharge standards.

- The cooling pond is suggested to be replaced by a thickener integrated with pretreatment to breakdown the complex substrates in POME, allowing anaerobic digestion to happen effectively and also as a means to regulate loading into the subsequent process;

Fig. 4.3 Proposed process flow for POME treatment

- The anaerobic ponds are proposed to be substituted by an "intelligent" high-rate anaerobic reactor with predictive and control capabilities for consistent effluent quality and biogas production; and
- The facultative/aerobic ponds are suggested to be substituted by a microbubble flotation aided with the coagulant process since the microbubble flotation was able to reduce BOD and other crucial discharge parameters below the stipulated discharge limits.

This concept would have reduced the footprint of POME treatment significantly when the area required to maintain the ponds will be eliminated. Furthermore, the implementation of the high-rate anaerobic reactor to replace the ponds could potentially reduce the emission of greenhouse gases contributing to global warming since the biogas produced will be collected for other purposes. Meanwhile, the use of microbubble flotation aided with coagulant is expected to be able to produce effluent that can be directly discharged to the watercourse and eliminating the need for an additional process to remove melanoidin (which gives POME the brown color) present in the wastewater. It could potentially provide palm oil mills a facelift of having an image of being the greatest polluters to be environmentally sustainable instead.

(ii) Possible advancements

Striving to continuously improve the POME treatment process, it is always essential to acknowledge flaws and identify possible areas for further development. Based on the concept presented in Fig. 4.3, the following are listed as areas with potential for further development:

(a) Substitution of polyaluminum chloride (PAC) as a material for coagulation in the microbubble flotation process: The agglomerated solids containing PAC cannot be directly disposed of as it is classified hazardous. It is important to handle it as a form of scheduled waste. To eliminate such problem, there is a need to search for a substituting natural coagulant that could work under similar conditions as PAC. The use of natural coagulants can potentially allow the use of the sludge after microbubble flotation to supplement as soil nutrients since the materials from POME are organic in nature.

(b) Producing value-added product from POME: Biogas can be a source to generate renewable energy. However, anaerobic digestion also has the potential to produce other value-added products such as volatile fatty acids (VFA) that could be raw materials to other processes. Currently, the production of VFA is limited by the condition of the anaerobic reactor. In order to produce VFA for extraction, two-stage anaerobic digestion can be explored as an alternative. The first stage involves hydrolysis and acidogenesis where a portion of VFA can be harvested. The effluent can then be supplied to the subsequent stage where acetogenesis and methanogenesis processes to generate biogas can be facilitated. There is still a gap in this area and a lot of fundamental work will be required to make the production of VFA possible from POME treatment.

(c) Co-digestion of POME with other milling wastes: Acknowledging that POME is not the sole source of waste that is generated from the palm oil milling process, the possibility of digesting solid wastes from palm oil mills (i.e., EFB) with POME can be explored. This could provide an alternative to the management of EFB should the focus of the mill is to produce sufficient biogas for power generation. Co-digestion of POME can also be coupled with other wastes in nearby industries. This leads to the possibility of a centralized wastewater treatment facility for various industries, reducing capital and operational costs.

While anaerobic digestion can do a lot to improve sustainability, another important issue to overcome will be the surplus of biogas. Currently, the palm oil mills are self-sufficient with the energy that they produce from the incineration of PKS and EFB. Producing more biogas without sufficient demand deters the industry from moving toward cleaner production.

Meanwhile, other energy-intensive industries and also rural areas that have an inconsistent supply of electricity do not have their energy demands met. This is attributed to the remote location and lack of access to connect the energy to the national grid. Should the electricity connection be provided, then palm oil mills will play a pivotal role to ensure the sustainability of energy generation to meet the demands of other industries, reducing reliance on fossil fuel. Moreover, this will also help develop more rural areas to improve their quality of life.

References

Atmaca, H., Cetisli, B., & Yavuz, H. S. (2001). The comparison of fuzzy inference systems and neural network approaches with ANFIS model for fuel consumption data. In *Second International Conference on Electrical and Electronics Engineering Papers ELECO '2001*. Bursa, Turkey.

Dellana, S. A., & West, D. (2009). *Environmental Modelling and Software, 24,* 96–106.

Khadaroo, S. N. B. A., Poh, P. E., Gouwanda, D., & Grassia, P. (2019a). Applicability of various pretreatment techniques to enhance the anaerobic digestion of Palm Oil MIll Effluent (POME): A review. *Journal of Environmental Chemical Engineering, 7,* 103310. https://doi.org/10.1016/j.jece.2019.103310.

Khadaroo, S. N. B. A., Grassia, P., Gouwanda, D., & Poh, P. E. (2019b). Is the dewatering of Palm Oil Mill Effluent (POME) feasible? Effect of temperature on POME's rheological properties and compressive behavior. *Chemical Engineering Science, 202,* 519–528.

Noori, P., Abdoli, M. A., Farokhnia, A., & Abbasi, M. (2009). Results uncertainty of solid waste generation forecasting by hybrid of wavelet transform-ANFIS and wavelet transform-neural network. *Expert System Application, 36,* 9991–9999.

Poh, P. E. (2012). *Treatment of palm oil mill effluent (POME) under thermophilic condition using upflow anaerobic sludge blankent-hollow centered packed bed (UASB-HCPB) reactor.* Ph.D. thesis submitted to the University of Nottingham in July 2011.

Poh, P. E., & Chong, M. F. (2014). Upflow anaerobic sludge blanket-hollow centered packed bed (UASB-HCPB) reactor for thermophilic palm oil mill effluent (POME) treatment. *Biomass and Bioenergy, 67,* 231–242.

Poh, P. E., Yong, W.-J., & Chong, M. F. (2010). Palm Oil MIll Effluent (POME) characteristics in high crop season and the applicability of high-rate anaerobic bioreactors for the treatment of POME. *Industrial and Engineering Chemistry Research, 49,* 11732–11740.

Tan, H. M., Poh, P. E., & Gouwanda, D. (2018). Resolving stability issue of thermophilic high-rate anaerobic palm oil mill effluent (POME) treatment via adaptive neural-fuzzy inference system predictive model. *Journal of Cleaner Production, 198,* 797–805.

Tan, H. M., Poh, P. E., & Gouwanda, D. (2019). *Automation of high-rate anaerobic reactor for palm oil mill effluent (POME) treatment for stable effluent and biogas production* (Unpublished).

Chapter 5
Sustainability of Waste Management Initiatives in Palm Oil Mills

5.1 Biomass from Oil Palm

Palm oil and palm kernel oil were the most traded oils in the world exports of oils and fats market with a market share of 59.6%, compared to soybean oil (15.3%), sunflower oil (7.1%), and rape-seed oil (5.2%) in 2010 (Oil World 2011). In 2018, Malaysia accounted for 39% of the world palm oil production and 44% of the world exports (http://www.mpoc.org.my/). While it facilitates the development of inclusive business involving strategic alliances between large-scale enterprises and small medium-size palm growers, and contributes to job creation and the development of the logistics, social and human infrastructure in the palm oil industry, those palm oil mills (POMs) that processed the Fresh Fruit Bunches ("FFB") contributed a significant amount of organic wastes and constituted the bulk of industrial solid wastes.

The oil palm waste production is at an average rate of 53 million tons annually with a 5% annual growth projection (Rezk et al. 2019), and it is projected to rise to approximately 100 million tons of oil palm biomass by the year 2020 (Umar et al. 2013; Mohammed et al. 2011). Nevertheless, due to the nonproductive and/or non-economical nature of the usual waste treatment mechanisms, the palm oil mills at times may be lax with compliance and enforcement. The laxity in compliance and neglect of proper waste treatment have become a cause for concern in palm oil production sustainability.

Most palm oil mills in Malaysia use the anaerobic ponding system for palm oil mills effluent (POME) treatment. The anaerobic ponding system would represent annual carbon emissions of more than 50,000 tons of carbon dioxide equivalent (tCO_2e) for a typical 45 MT per hour capacity palm oil mill. If the anaerobic ponding system is not maintained appropriately, the noncompliance discharge from effluent ponds coupled with significant chemical fertilizers runoff (due to hardening of soil conditions caused by long-term intensive chemical fertilizers applications) would represent the major source of pollutants in the waterways in the farming communities. The resulting environmental issues are alarming.

© Springer Nature Switzerland AG 2020
P. E. Poh et al., *Waste Management in the Palm Oil Industry*,
Green Energy and Technology, https://doi.org/10.1007/978-3-030-39550-6_5

A typical operating palm oil mill would produce the following waste streams consisting of the following:

- Empty fruit bunches (EFB)—23% of FFB;
- Potash ash—0.5% of FFB;
- Palm kernel—6% of FFB;
- Fiber—13.5% of FFB; and
- Shell—5.5% of FFB.
- Palm Oil Mill Effluent—80% of FFB
- Decanter Sludge—3% of FFB

(Source: Ministry of Science, Technology and the Environment Malaysia 1999).

5.2 Our Concerns in Sustainability

Malaysia is one of the countries pledged to cut carbon dioxide emission intensity by 45% by 2030, which is in line with the United Nation's 2030 Agenda for Sustainable Development Goal that calls for urgent action to combat climate change and its impacts. The palm oil industry represents one of the industries that contribute to carbon emissions in Malaysia. In the event the current palm oil mills continue to employ anaerobic ponding system for POME treatment, it would represent potentially 35,000–60,000 tCO_2e per year per palm oil mill. The achievement of this ambitious goal requires a change in attitude on the part of the producers and consumers in the economy. Such a change has the potential to influence them to engage in environmentally friendly practices.

There are concerns that affect long-term sustainability of the palm oil production predominately with regards to soil ecosystem degradation, long-term intensive chemical fertilizers applications, removal of biomass from the estates/plantation, and nonproductive waste treatment mechanisms, which ultimately lead to yield decline (Fig. 5.1).

The excessive use of chemical fertilizers in palm oil production has caused land degradation (especially for the second- and third-generation replanting estates), deteriorated soil conditions of most farms, and eventually decreased yield and oil extraction rate. This has made second- and third-generation replanting estates uneconomical and unsustainable. In addition, the degrading soil condition has affected the soil ecosystem; a significant amount of the applied chemical fertilizers was not absorbed by the palm trees and leached away to the waterways that lead to pollution of water sources next to the plantations.

One of the concerns affecting the yield in palm oil production is the Ganoderma diseases, caused by the white rot fungus Ganoderma boninense (Flood et al. 2000). Basal stem rot (BSR) is the most common manifestation of Ganoderma disease in the region (Malaysian Palm Oil Board 2019) representing potentially the greatest threat to sustainable oil palm production in South East Asia (Flood et al. 2010). As the pathogen is not a good competitor within a healthy soil ecosystem, allowing the

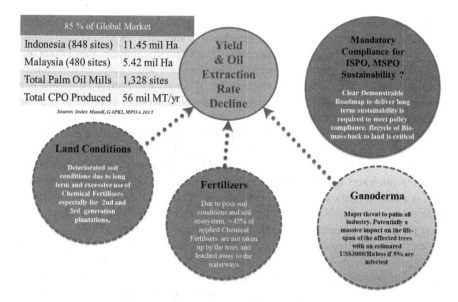

Fig. 5.1 Concerns in achieving real sustainability

natural microbial population to reduce the effective inoculum augurs well for long-term soil remediation that improves soil ecosystem and should be considered as a management option for this disease.

5.3 Meeting Sustainable Palm Oil Policies Compliance

Industry players are under pressure to achieve the Sustainable Palm Oil Certificate, Roundtable on Sustainable Palm Oil (RSPO) as well as to ensure that the environment and the rights and well-being of the local commodities and indigenous groups are preserved. Besides RSPO, both the Malaysian and Indonesian governments' sustainability certification initiatives for palm oil have emerged. Among those are the mandatory Malaysian Sustainable Palm Oil (MSPO) certification, and sustainable production of all biofuel feedstocks, including oil palm biomass, can be certified by these sustainability certifications.

Today, the application of green technology is the key contributor to preserve the long-term sustainability necessary to protect renewable resources. To increase export capacity and improve market compliance, there should be greater efforts to reduce GHG emissions by promoting sustainable farm management. And the adoption of the MSPO certification is made mandatory among industry players to mitigate adverse effects on local palm oil production.

The stakeholders are very much concerned about how to achieve the real sustainable palm oil production and meet the mandatory compliance for MSPO sustainability while improving productivity and profits. To address this concern, the optimal use of biomass waste from palm oil industry functions as a waste disposal mechanism is salutary since the number of wastes produced is massive. Fortified compost is a key farm input to remediate and revitalize the soil ecosystem in order to provide sustainable yield improvement.

Oil palm in plantations is an energy-efficient crop that requires the least energy input to produce a ton of oil vis-a-vis other vegetable oils. Malaysia is the highest consumer of inorganic fertilizer per hectare, and imports more than 90% of inorganic chemical fertilizers and 60% of the cost of production in oil palm is on fertilizers. This is a major challenge as the increased use of chemical fertilizers in the large-scale palm oil plantations has degraded the soil and the resulting decreased yields exert pressure to open up more land for cultivation. The organic EFB waste from POMs that are left to decay in the landfills and liquid waste from POME would emit various Greenhouse Gasses (GHG). Opening up new land for cultivation will exacerbate pollution and climate change. For these reasons, the sustainability of palm oil production has always been questioned. Long-term sustainable methods are a critical and a fundamental change in the processes which should be put in place to recycle organic waste from POMs and plantations.

At the time of writing this article in Feb 2019, The European Commission has published its proposed criteria for determining what crops caused harm, following a law passed by the European Union in 2018 to end the use of feedstocks in biofuels that damage the environment.

Under the new EU law, the use of more harmful biofuels will be capped at 2019 levels until 2023 and reduced to zero by 2030. However, it also provides that producers who could show they had intensified yields may be exempt. It could then be argued that their crops cover demand for biofuel and for food and feed, without needing expansion onto nonagricultural lands, such as forests (Blenkinsop 2019).

The merit of the biomass recycle approach is that it processes or converts all organic wastes from palm oil mill into specific blends of fortified compost and/or Bioorganic Fertilizers ("BOF") which are redeployed to the surrounding communities and plantations for long-term soil remediation. This biomass recycle approach represents a very significant improvement in the reduction of pollutants from the usual waste treatment process and has, besides the environmental, inherent economic values that compel the stakeholders to adopt and maintain it. The application of appropriate fortified compost or BOF as integral supplements in addition to chemical fertilizers (in reduced amount) would form the basis of a sustainable plantation management practice that promotes better plant health, higher fruits yield, and prolongs the productive lifespan of the biological assets.

5.4   Waste Treatment Options and Considerations

In the past, palm oil industries have adopted various waste treatment mechanism for different organic waste streams, i.e., EFB (pressed or unpressed), POME, decanter sludge/cake, boiler ashes, kernel shells, and palm mesocarp fibers from the palm oil mills. Apart from kernel shells and palm mesocarp fibers which are predominately redeployed as palm oil mill own biomass power generation feedstocks, each of the various waste treatments mechanism has their pros and cons as well as different economic values for sustainability (see Fig. 5.2).

Among various organic wastes streams, there are two major waste streams: the landfills for solid organic waste (such as EFB) and anaerobic ponding system for liquid waste POME. These two waste streams contribute the most to the GHG emissions.

Over the years, most palm oil plantations have considered using or processing the solid waste, EFB into composts, raw fiber materials for industrial products or biomass for power generation. All these mechanisms have various degrees of complexity, logistics, and economic challenges.

The optimal use of biomass waste can be achieved through waste treatment mechanisms that deliver higher productivity and/or additional economic benefits in addition to being environmentally friendly. However, some of the distinct waste treatment processes for different waste streams adopted (see Fig. 5.2) would, instead of adding value, add operational costs to the industry.

In addition to the operational and economic challenges, if the biomass produced by the estates are removed without being recycled to the field, then the waste treatment mechanism adopted causes discontinuity in sustainability in palm oil production.

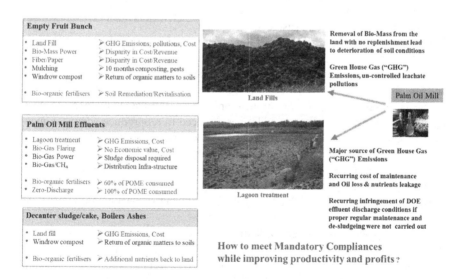

Fig. 5.2 Waste treatment options

The long-term removal of these biomasses which deprive the land of the necessary organic matters for healthy soil ecosystems would eventually lead to land degradation. A healthy soil ecosystem is one of the fundamental factors to improve yield and productivity.

Treating biomass waste with advanced waste treatment technologies that recycle the biomass back to the land will lead to sustainability intensification. For this, the biomass can be used to produce BOF for palm oil plantations. The reapplication of BOF onto farmland for soil remediation can subsequently reduce the reliance on chemical fertilizers. It promotes not only long-term soil remediation coupled with a nutrient replacement but also potentially become the new standard for sustainable husbandry, which will contribute to long-term sustainability. Further, the CO_2 emissions are reduced during the processing of the biomass waste.

Over the years, attempts to recycle the EFB in the form of mulch and/or compost have been frustrated by the absence of a well-established mechanism that can clearly demonstrate the cost and benefits of such approaches. Therefore, it is essential to search for a sustainable solution to provide long-term soil remediation to the palm oil plantations without harming the environment, while improving the sustainability of the palm oil industry. The mechanism that provides the optimal biomass utilization in terms of economic and environmental values should be identified.

5.5 Bioorganic Fertilizer (BOF) and Bio-circular Economy

In the palm oil industry, the recycling of biomass as a fortified compost/BOF is referred to as a Bio-circular economy principle framework. This framework is significant for developing countries that are contemplating rolling out large-scale plantations in a practical and sustainable manner.

The key priority that is to achieve long-term sustainability is having soil remediation as a critical fundamental process, i.e., the recycling of organic waste from POM should be considered as the primary process between the farm and the mill (see Fig. 5.3).

The agronomic principal in the application of these BOFs as supplements to existing chemical fertilization regime addresses the need to reintroduce adequate quantities of beneficial soil-borne microbes to counter the negative effects of prolonged use of chemical fertilizer (Glick 1995; David and King 2005; Parr et al. 1994). The consistent applications of these BOFs in addition to chemical fertilizers (in reduced amount) represent a paradigm shift in the usual palm oil plantation management. These sustainable practices have established significant real economic returns and benefits.

Putting in place an integrated waste treatment plant, next to or near a palm oil mill that processes all organic waste streams, liquid, and sludge discharges from biogas plants (If any) into fortified composts such as BOF, would represent an essential infrastructure building for sustainable palm oil production. A sustainable and consistent method of integrated waste treatment would help to minimize environmental

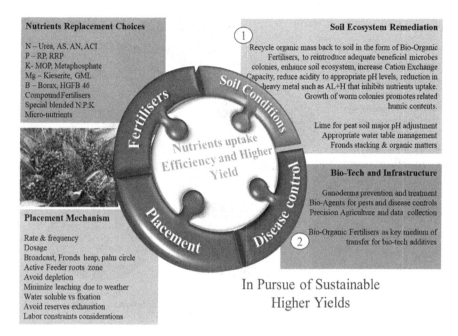

Nutrients Replacement Choices

N – Urea, AS, AN, ACI
P – RP, RRP
K- MOP, Metaphosphate
Mg – Kieserite, GML
B – Borax, HGFB 46
Compound Fertilisers
Special blended N:P:K
Micro-nutrients

Placement Mechanism

Rate & frequency
Dosage
Broadcast, Fronds heap, palm circle
Active Feeder roots zone
Avoid depletion
Minimize leaching due to weather
Water soluble vs fixation
Avoid reserves exhaustion
Labor constraints considerations

Soil Ecosystem Remediation

Recycle organic mass back to soil in the form of Bio-Organic
Fertilisers, to reintroduce adequate beneficial microbes
colonies, enhance soil ecosystem, increase Cation Exchange
Capacity, reduce acidity to appropriate pH levels, reduction in
heavy metal such as AL+H that inhibits nutrients uptake.
Growth of worm colonies promotes related
humic contents.

Lime for peat soil major pH adjustment
Appropriate water table management
Fronds stacking & organic matters

Bio-Tech and Infrastructure

Ganoderma prevention and treatment
Bio-Agents for pests and disease controls
Precision Agriculture and data collection

Bio-Organic Fertilisers as key medium of
transfer for bio-tech additives

Nutrients uptake Efficiency and Higher Yield

Soil Conditions

Fertilisers

Placement

Disease control

In Pursue of Sustainable
Higher Yields

Fig. 5.3 Factors in sustainable yield intensification

impact. Eventually, it would enhance the country's energy regime into a more sustainable level (Fig. 5.4). The precision agriculture platform and services within any Public–Private Partnership Bio-Circular Economic Framework could be/should be extended to the smallholders to better manage their estates.

5.6 Case Study on Bio-circular Economic Framework

We will discuss a particular palm oil mill that has adopted a Bio-Circular Economic Framework business to treat and convert palm oil mill wastes into BOF in the state of Sabah Malaysia from 2009–2018. (UNFCCC CDM Project—Tingkayu Palm Oil Mill 2006: https://cdm.unfccc.int/Projects/Validation/DB/HFPCBID1DOEN7SOUQCUY9F812668X9/view.html).

5.7 Waste Returned to the Plantation for Mulching Versus Fortified Composts

The utilization of EFB for mulching requires access to plantations and carries a high cost associated with the transportation and application of mulched EFB on land.

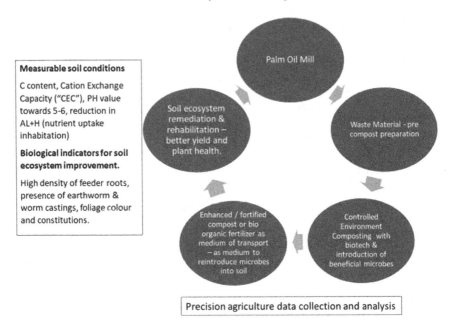

Measurable soil conditions

C content, Cation Exchange Capacity ("CEC"), PH value towards 5-6, reduction in AL+H (nutrient uptake inhabitation)

Biological indicators for soil ecosystem improvement.

High density of feeder roots, presence of earthworm & worm castings, foliage colour and constitutions.

Palm Oil Mill

Soil ecosystem remediation & rehabilitation – better yield and plant health.

Waste Material - pre compost preparation

Enhanced / fortified compost or bio organic fertilizer as medium of transport – as medium to reintroduce microbes into soil

Controlled Environment Composting with biotech & introduction of beneficial microbes

Precision agriculture data collection and analysis

Fig. 5.4 Bio-circular economic framework

Many plantations do not use EFBs for mulching, as it has limited fertilizer value per ton of mulch applied, can remove nitrogen from the soil, and the cost of application/distribution into the estate is relatively high as the volume to be mulched and distributed on a daily basis is substantial, especially, in the face of the shortage of labor. It can also spread plant diseases which are a constant source of concern for all palm oil growers and encourages the propagation of rhinoceros beetles, which damage the palm trees (Kamarudin and Wahid 2004). Thus, the distribution of mulching also faces another barrier, as it has a detrimental effect on the plantations (Menon et al. 2003). It is not considered as a plausible better alternative to fortified composts.

Using the operational data collected, we will be able to extrapolate the economic, social, and environmental benefits as direct ramifications of BOF production activities of the deployment of such bio-circular economic framework for the palm oil industries. The two main technologies employed were (1) In-Vessel biomass composting and (2) Biotechnology with the introduction of "Environmentally friendly Indigenous Beneficial Microbes to promote soil remediation through intensifying microbes".

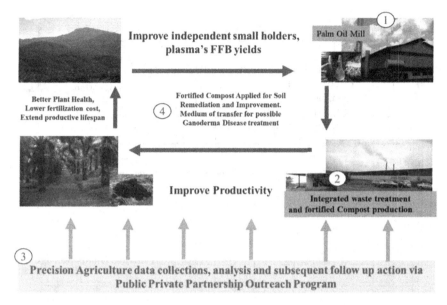

Fig. 5.5 Case study: integrated waste treatment and bioorganic fertilizers production plant

5.8 In-vessel Anaerobic Composting—Control Environment Composting Chambers ("CECC")

The case study project activity involved the construction and operation of a waste treatment plant next to the existing palm oil mill, thereby reducing the cost of transporting the waste materials used in the treatment process (see Fig. 5.5).

The waste treatment plant will consist of a pre-composting section, in-vessel composting, and post-composting sections with a total area of approximately 0.8 ha. The plant layout is shown in the following Fig. 5.6.

5.9 Wastewater Treatment Process

Another key characteristic of these types of enhanced anaerobic composting mechanisms under controlled environment is the utilization of the POME as the source of fluid required and form part and parcel of the highly aerated composting process. The suspended solids/organic matters present in the POME form part of the ingredients for the production of BOF. The raw POME discharge from the palm oil mill was injected into a composting vessel in the automated process to maintain constant moisture content in the composting feedstock. As the In-Vessel composting process, at full capacity, consumed some 60% of the POME, the remaining POME was treated in anaerobic lagoons. In the anaerobic lagoons, the bulk of the remaining

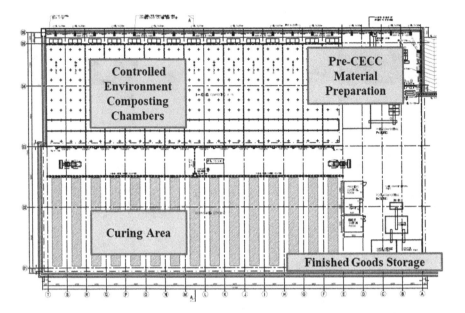

Fig. 5.6 Case study: waste treatment plant layout

BOD/COD in the POME is converted to CO_2 without the formation of methane. The entire process of treating POME under anaerobic composting conditions takes place within the CECC. The excess leachate from CECC is treated in the aerated ponding system. The treatment plant process and follow up anaerobic treatment essentially treats all of the POME effluents under anaerobic conditions.

5.10 In-vessel Biomass Composting

The waste EFBs from the palm oil mill will be shredded to a predetermined size in the pre-composting section and then mixed with sludge and boiler ash. The resulting biomass mixture will then be transferred by front-end loader into the CECC and piled into heaps of approx. 2 m in height along the full width (6 m) and length (25 m) of the tunnels. Once filled, the tunnel will be sealed off and the air blowers start under computer control to ensure optimum composting process conditions according to a predetermined batch processing temperature schedule. The composting process is activated using a mixture of specially formulated microorganisms mixed with the POME and sprayed on to the composting heap. The computer managed control system will enable completion of the composting process in the vessel in 2 weeks by careful control of oxygen levels in the biomass as well as the temperature and moisture levels. At the high temperatures in the process, large quantities of POME are sprayed onto the composting waste and evaporated, thereby maintaining the optimum

Controlled Environment Composting
- Seal & computer controlled Composting Chambers
- Optimum composting conditions 7 X 24, via high aeration, with 14 days batch cycle
- POME application for moisture control & Bio-Formulation

- Direct delivery of waste streams into the Plant
- Jointly operated with palm oil mill
- Option for Zero Discharge
- Option: As upgrade for existing windrow composting
- Option: In-situ Bio-Refinery est. RM12 mil
- Option to integrate with Bio-Gas power for self sustaining Power Generation or/and Compressed CH_4 production

Fig. 5.7 Controlled environment composting chambers

moisture levels. After 2 weeks, the composted waste will be removed by a front-end loader from the vessel and matured in piles for a further 14 days to allow the curing process to continue and temperature to reduce by natural cooling (see Fig. 5.7).

The In-vessel composting system deployed when compared to the usual windrow-based composting mechanism addressed the major hindrance and shortfalls of the conventional windrow composting system especially in terms of consistency in the production of bio-formulation fortified composts/BOF.

The efficacy of these BOF with consistent high CFU presence (CFU > 10^{-9} observed in the case study) of beneficial microbes would enable the planters to observe the key performance indicators in terms of soil conditions, plant health, and yield increase when compared to unfortified composts produced from windrow composting mechanism (see Fig. 5.8). In the uncontrolled environment of the usual windrow composting process, the composting period varies from 45 days to 60 days in a covered/sheltered windrow composting plant. To ensure consistent quality of the bio-fortified compost/BOF is an operational challenge. It prevents determining the efficacy of the usual windrow-based compost when applied to the fields.

As all the organic waste streams are composted under the anaerobic environment, similar integrated waste treatment and BOF production plant would be able to establish the total emission reductions of the project activities over the first 10 years. This is expected upon full production to be 61,050 tCO2e (see Table 5.1).

	Outdoor system	Indoor System	CECC
Composting site	Open air > 10 Ha	Sheltered indoor > 4 Ha	Controlled enclosed environment - 1 Ha
Weather protection	Compost cover, No rain water run off control	Shelter/roofing	Controlled indoor
Operational complexity. Composting period	Manual supervision 60-90 Days	Manual supervision 45-60 Days	Computer Controlled 14 Days
Additional moisture	Uncontrolled (weather)	Not fully controlled	Fully controlled environment
Temperature Control	No	No	Fully controlled environment
Leachate control	Minimum	Full	Controlled and recyclable
Quality Control & Traceability	No	Partial Compliance	Full Compliance
Production of Bio-organic Fertilisers	Non-Compliance Low quality compost only	Partial Compliance Inconsistent Quality & CFU	Full Compliance High CFU count

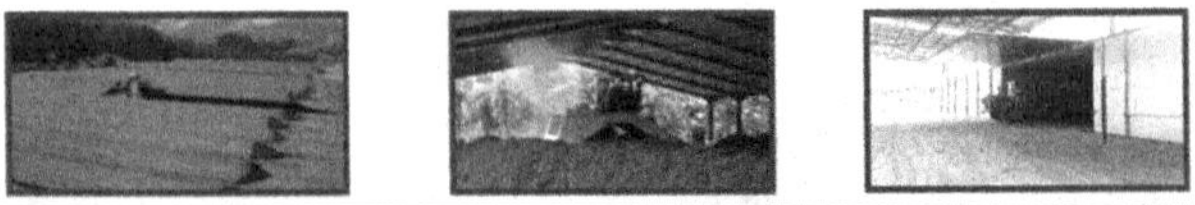

Fig. 5.8 Windrow-based systems versus CECC

Table 5.1 Annual estimation of emission reductions

Years	Annual estimation of emission reductions in tons of CO2e
Year 1	32,392
Year 2	41,781
Year 3	49,702
Year 4	56,384
Year 5	62,022
Year 6	66,778
Year 7	70,791
Year 8	74,177
Year 9	77,033
Year 10	79,443
Total	610,502
Total number of crediting years	10 years
The annual average of estimated reductions (tons of CO2e)	61,050

5.11 Key Economic Deliverables for a Typical 45 MT/hr Capacity Palm Oil Mill

A typical 45 MT/hr capacity palm oil mill would process approximately 200,000 MT of FFB collected from the surrounding oil palm estates of 10,000 ha (avg. 20 MT/ha/year for matured tree)—12,500 ha (avg. 16 MT/ha/year for 25-year-old tree), depending on the FFB yield per ha. The expected organic waste streams will be as shown in Table 5.2.

Based on the operational and economic data from the case study, apart from the obvious environmental benefits in recycling the biomass as fortified compost/BOF and meeting the Mandatory Sustainable Palm Oil policy of Malaysia and Indonesia, the key economic benefits can be observed in the following segments (see Table 5.3) subject to site locations and local conditions.

Table 5.2 Typical major solid and liquid wastes volume for 45 MT/hr capacity palm oil mill

	Average per year
Fresh fruits bunches processed by palm oil mill (45 Mt/hr)	200,000 MT
Waste streams	
Empty fruit bunches	42,000 MT
POME	160,000 M^3
Decanter sludge	9,000 MT
End products	
Bio-organic fertilizers	14,700 MT
Certified emission reductions	30,000–60,000 tCO2e, *subject to baseline conditions*

Table 5.3 Key economic and environmental benefits

Benefits[a] Total: RM12.61 mil/year/site	RM per year[a]
Waste treatment productivity gains (including EFB supply + Power cost recovery via sales of BOF)	980,000
Logistics saving	930,000
Overall fertilization cost savings with reducing chemical fertilization rate	1.5 mil
Yield increase of minimum of 2 MT/Ha/year	**9.2 mil**
Certified emissions reductions	30,000–60,000 tCO2e

[a] Assuming OER of 0.20 and CPO price of RM2,300/MT

5.12 Waste Treatment Productivity Gains

The cost of POME treatment by traditional anaerobic lagoon/ponding system includes the yearly ponds sediment desludging and the maintenance of polishing plant before discharge is approximately RM2.6/m^3. This would represent approximately RM416,000/annum of saving when the POME is consumed by the CECC. The recovery of the supply of EFB and power as the cost of the BOF feedstock and utilities based on the case study is RM38.5/MT of BOF produced which represents a productivity gain of RM565,950/annum. Total extrapolated waste treatment productivity gain is RM981,950/annum.

(i) Logistics savings

In the business as a usual scenario without Bio-Circular economic approach, each typical POM would incur the cost of EFB removal from POM. The average cost of removal is RM15/MT. During normal POM operations, the trucks that deliver the FFB to the palm oil mill returned to the estates empty. In the case study, the FFB delivery trucks transport the BOF back to the plantation as and when the BOF was ready for shipment which would represent cost saving of RM30/MT of BOF delivered back. The total logistic cost savings would be RM630,000 (EFB) + RM300,000 (BOF, assuming BOF application rate of 1 MT of BOF per ha per year for 10,000 ha).

(ii) Overall fertilization cost savings with reduced chemical fertilization rate

Depending on the surrounding plantation soil conditions and nutrient deficiency profile, the usual chemical fertilizers application rate will be retuned on the general approach of reducing chemical fertilizers rate supplemented with BOF for soil remediation. Better soil conditions typically include improvement in soil Cation Exchange Capacity, the lowering of soil acidity toward PH 7, and improvement in soil texture (loosening) and water retention capacity. With better soil conditions, the palm trees would have a better nutrient absorption rate. Given these positive changes, the revised/retuned fertilization protocol would be toward a cost saving target of RM150/Ha per year. For a deployment of 10,000 ha, it would represent a yearly saving of RM1.5 mil.

(iii) Yield increase of a minimum of 2 MT/Ha/year

Given the improved soil conditions via the BOF+Chemical fertilizers regime in the case study and observed yield increase of minimum 2 MT of FFB/ha. When compared to negative control plots of the previous full chemical fertilizers regime, this would represent an economic gain of RM9.2 mil when 10,000 ha of palm oil estates participate in this Bio-Circular economic approach.

Extrapolating from the above economic benefits indicators from a typical 45 MT/hr palm oil mill, the Malaysian palm oil industry stands to gain a significant bio-circular economic increase of RM6B per year. (480 POMs × RM12.61 mil increase per years).

In terms of ROI, this bio-circular economic approach would represent the most optimal waste treatment option when compared to other waste treatment mechanism, not including the long-term sustainable yield improvement in palm oil production. The pressure to open new land for palm oil cultivation to replace the nonproductive old palm oil estates would be reduced.

(iv) Opportunities for improvement

With the implementation of the Malaysian Government Economic Transformation Program in early 2011, there are 19 Entry Point Projects (EPPs) and developments that fulfill the 10 National Key Economic Area (NKEA). Of these, there are eight core EPPs that span the palm oil value chain, such as accelerating the replanting of oil palm, improving fresh fruit bunch (FFB) yield, improving worker productivity, increasing oil extraction rate, and developing biogas at POMs. EPPs aim at promoting the recycling of POM wastes back to plantations in various blends of BOF targeted at improving plant health and yields.

A CECC based Bio-Circular Economy approach can be a foundation for good plantation management practice. It recycles a significant amount of organic matters fortified with beneficial microbes back to land with efficient logistical arrangement by utilizing empty FFB transport vehicle to ship BOF back to the plantation. It remediates soil conditions to improve cation exchange capacity which lead to better absorption of nutrients (NPKs) and provides timely fertilization schedule to take into account the effect of climate changes, the production capacity of POM, and land left fallow, apart from scheduling timely balancing of harvest process and applications of fertilizers.

Fortified compost/BOF can be utilized as a delivery medium for additional bio-agents or bio-formulations for plant disease control, in particular in the management of Ganoderma disease where potentially healthy soil ecosystem could reduce the infection rate. More studies are required to establish this correlation.

There have been many types of research done on the potential of using various biotech or microbial formulation to improve plant health yields and disease control such as Ganoderma disease control, (MPOB TS 53 2008; MPOB TT 474, 2010; MPOB TT 443, 2010).

These BOF generated under the Bio-Circular economy approach can be used as the key medium for the delivery of various bio-formulations or bio-agents by plantations or Industry R&D units. The Bio-Circular Economy approach potentially can represent one of the go-to-market mechanisms in the commercialization of these new biotechnologies. The cost of introducing the additional formulation is incremental, and the usual implementation challenges such as logistics and capacity building can be overcome via existing BOF production facilities, BOF offtake, and deployment infrastructure.

Within the Bio-Circular Economy framework, there can be potential Public–Private Partnership provisions for open cooperation model whereby joint cooperation mechanism with various government agencies, small and medium enterprise/technology companies, and the plantation groups can be facilitated to leverage off on existing CECC sites/plants to exploit this sustainable resource management opportunity.

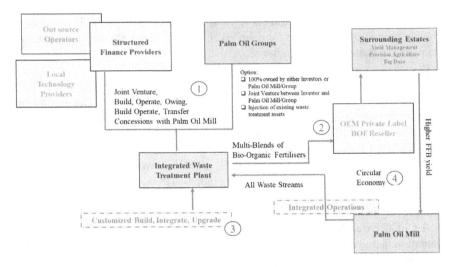

Fig. 5.9 Potential Public–Private Partnership initiative model

5.13 Conclusion

This study shows that it is feasible and more cost-effective to use organic bio-fertilizer. This approach offers several environmental and cost advantages.

Moving Forward

There is a need to redefine the role of palm oil mill as the nucleus for the upstream communities especially as the center of knowledge dissemination on various Best Agricultural Practices (BAPs) that are endorsed and supported by government agencies and policy makers. Using proven operational KPI's from existing BAP control environmental composting practitioners to put in place sustainable share economic model for the deployment of infrastructure for long-term soil remediation and adoption of this BAP for the smallholder's communities. An inclusive Public–Private Partnership Program is required to kick start this significant BAP infrastructure to bridge the sustainability gap for the smallholders, plasma–plasma, clusters and improve their income and livelihood (see Fig. 5.9).

References

Blenkinsop, P. (2019). *EU targets palm oil for road fuel phase-out, but with exemptions*. Reuters. Retrieved February 12, from https://in.reuters.com/article/us-eu-biofuels/eu-targets-palm-oil-for-road-fuel-phase-out-but-with-exemptions-idINKCN1Q021Q.

David, S., & King, C. (2005). *Affect of improving soil organic matter with compost on broad-acre crop production. Organic farming systems*. http://primalfoodsgroup.com/wp-content/uploads/2015/05/2004-Broad-Acre-crop-production.pdf.

Flood, J., Bridge, P. D., & Holderness, M. (Eds.). (2000). *Ganoderma diseases of perennial crops* (275 pp.). Wallingford, UK: CABI Publishing.

Flood, J., Cooper, R., Rees, R., Potter, U., & Hasan, Y. (2010). *Some latest R&D on ganoderma diseases in oil palm.* https://assets.publishing.service.gov.uk/media/57a09de540f0b64974001ae0/Ganoderma-Diseases.pdf.

Glick, B. R. (1995). The enhancement of plant growth promotion by free living bacterial. *Canadian Journal of Microbiology, 41,* 109–117.

Kamarudin, N., & Wahid, M. B. (2004). Immigration and activity of Oryctes rhinoceros within a small oil palm replanting area. *Journal of Palm Oil Research, 16*(2), 64–77.

Malaysian Palm Oil Board. (2019). *Plant protection industry.* Retrieved February 25, 2019, from http://www.mpob.gov.my/ms/component/content/article/734-plant-protection-industry.

Mohammed, M. A. A., Salmiaton, A., Wan Azlina, W. A. K. G., Mohammad Amran, M. S., Fakhru'l-Razi, A., & Taufiq-Yap, Y. H. (2011). Hydrogen rich gas from oil palm biomass as a potential source of renewable energy in Malaysia. *Renewable and Sustainable Energy Reviews, 15*(2), 1258–1270.

Menon, N. R., Zulkifli A. R., & Nasrin A. B. (2003). Empty fruit bunches evaluation: Mulch in plantation vs. fuel for electricity generation. *Oil Palm Industry Economic Journal, 3*(2), 15–20.

Ministry of Science, Technology and the Environment Malaysia. (1999). *Industrial processes and the environment, handbook no. 3.* Crude Palm Oil Industry, Department of Environment.

Oil World. (2011, December). *Oil world annual vol 1—up to 2010–2011.* Hamburg: ISTA Mielke GmbH.

Parr, J. F., Hornick, S. B., & Kaufman, D. D. (1994). *Use of microbial Inoculants and organic fertilizers in agricultural production* (16 pp.). xtension Bulletin No. 394. Taipei, Taiwan: Food and Fertilizer Technology Center (FFTC). http://www.fftc.agnet.org/htmlarea_file/library/20110722114739/eb394.pdf.

Rezk, H., Nassef, A. M., Inayat, A., Sayed, E. T., Shahbaz, M., & Olabi, A. G. (2019). Improving the environmental impact of palm kernel shell through maximizing its production of hydrogen and syngas using advanced artificial intelligence. *Science of the Total Environment, 658,* 1150–1160.

Umar, M. S., Jennings, P., & Urmee, T. (2013). Strengthening the palm oil biomass renewable energy industry in Malaysia. *Renewable Energy, 60,* 107–115.

UNFCCC CDM Project—Tingkayu Palm Oil Mill. (2006). https://cdm.unfccc.int/Projects/Validation/DB/HFPCBID1DOEN7SOUQCUY9F812668X9/view.html.

Printed in the United States
By Bookmasters